Coastal Geomorphology of India

Shrikant Karlekar

Diamond Publications

Coastal Geomorphology of India

Shrikant Karlekar

First Edition : December, 2016

ISBN : 978-81-8483-652-3

© **Diamond Publications, 2016**

We gratefully acknowledge the financial assistance from
Goa State Biodiversity Board, Saligao, Goa
for the publication of the book titled **Coastal Geomorphology Of India.**

Cover Page
Sham Bhalekar

Published by
Diamond Publications
264/3 Shaniwar Peth, 302 Anugrah Apartment
Near Omkareshwar Temple, Pune - 411 030
☎ 020-24452387, 24466642
info@diamondbookspune.com

For Online Shopping Visit to
www.diamondbookspune.com

Sole Distributor
Diamond Book Depot
661 Narayan Peth
Appa Balwant Chowk
Pune 411 030
Tel. - 24480677

CONTENTS

FOREWORD

'Coastal Geomorphology of India' authored by Dr. Shrikant Karlekar is a concise geomorphic account of Indian coasts, with a discussion of underlying processes and the resultant landforms in their specific regional and environmental context. The book covers the significant offshore feature continental shelf, the shoreline and the coast, extending from the Rann of Kutchh, on the west coast of India, to the Ganga delta in West Bengal and the coastline of the island state of Andamans and Nicobar in the Bay of Bengal. The analysis and discussion of the coastal geomorphology follows a state-wise sequence, starting from Gujarat, Karnataka and Kerala, Tamil Nadu, Andhra Pradesh, Odisha, and ending with West Bengal , with an addition of the Bay islands.

Having specialized in coastal geomorphology, Dr. Karlekar has devoted several decades of his scholarly pursuit to the study of the west coast of India. especially the Konkan coast of Maharashtra. This book is the extension of many of his publications on the west coast, with a fair amount of input from secondary sources and the research findings of many contemporary scientists. Divided into five chapters, with a preface and an introduction, the main focus of study is the geomorphology of the west and east coast of India, discussed in chapter four and five respectively: the former, the area of author's intensive study, occupies the largest space in the book.

The contents of the chapters are well illustrated with maps, diagrams and coloured satellite images, superimposed with clear delineation of landforms to facilitate their identification. The book carries a select bibliography that will prove valuable to those engaged in coastal studies and especially coastal geomorphology of India.

This volume is a helpful text to initiate those interested in the study of Indian coasts, to some of its basic features and characteristics.

K. R. Dikshit
(Former Professor and Head, Dept. of Geography,
Savitribai Phule Pune University, Pune)

PREFACE

A necessity to understand the processes operating along the Indian coasts is being increasingly felt by the researchers for the proper management of the coast. This has warranted an in-depth study of sections of our coast by field visits, field mapping, field observations and measurements and interpretations from aerial photographs and satellite images of the coast. This has resulted into many welcome additions to the literature on coastal geomorphology of India that have certainly ameliorated our understanding of the forms and processes on the Indian coast.

It can be seen that most of the research articles and research papers published in various books and journals discuss the landforms and analyze the processes along certain segments of the Indian coast or at certain specific sites. These placed together no doubt give an integrated picture of what the Indian coast is like geomorphologically.

The coastline of India comprises of headlands, promontories, rocky shores, sandy spits, barrier beaches, open beaches, embayments, estuaries, inlets, bays, marshy lands, deltas and offshore islands. These coastal and shoreline environments have been strongly affected by sea level changes throughout the geological history.

This book on 'Coastal Geomorphology of India' is an account of coastal processes and forms developed along the coastal stretch of every state of India. The West coast of India is geomorphologically different from the East coast in many respects. Segmental study of our coast brings out these differences very clearly and tells us why and how the morphology of the coasts varies from place to place.

The book includes discussions on major aspects of Geomorphology of Indian coast beginning from the nature of continental shelf, sediments on Continental shelf and continues with the description of coastal climate, coastal processes and landforms with reference to the coastal segment of each state. It is based on the field studies, interpretations of aerial photographs and Google earth and satellite images of the coast. The book draws extensively on the findings reported in various research articles and research papers published in books and journals devoted to coastal research. In some cases it also tries to examine temporal aspect of landform development within Holocene period.

It is hoped that this volume will serve as a useful source book for researchers, students and managers of coastal areas in every coastal state of India.

Appropriate acknowledgement is made in the book where the illustrations are used from other sources. I take this opportunity to thank Mr. Dattatraya Pashte, the publisher of Diamond Publications, Pune, for his willingness and support in the publication of volume.

Shrikant Karlekar

V

ACKNOWLEDGEMENT FOR ILLUSTRATIONS / FIGURES AND TABLES

Figure No.	Journal / Institute
3	Journal of Geological Society of India, Vol.46 (1995)
4	Journal of Geological Society of India, Vol. 55(2001)
6	National Institute of Oceanography (CSIR / NIO)
12	Journal of Coastal Conservation, Vol. 12 (2008)
13	National Institute of Oceanography ,2003
15	Journal of Earth System Science, 123 (1), (2014)
21	Integrated Coastal and Marine Area Management,(1992)
51	Indian Journal of Marine Sciences, Vol.30 (4),2001)
66	Indian Geographical Journal, Vol.60, (1985)
70	Ind. Journal of Geography and Environment Vol.13 (2014)
Table 1	Centre for Coastal Zone Management and Coastal Shatter Belt
Table 2	Current Science, Vol. 91, No. 4
Table 3	Current Science, Vol. 91, No. 4
Table 4	Current Science, Vol. 91, No. 4
Table 5	Current Science, Vol. 91, No. 4
Table 6	Current Science, Vol. 91, No. 4
Table 24	Atlas of Mangrove Wetlands of India

INTRODUCTION

India has a 7517 km long stretch of coastline (5423 km along the mainland and 2094 km along the Andaman, Nicobar and Lakshadweep islands), a coastline that is very complex as regards coastal landforms and the coastal processes (Fig 1). Despite such a long coastline, coastal studies in India were only a few till last decade or so. Past few years have witnessed a spurt in research in the field of coastal geomorphology. There are many welcome additions to the literature on coastal geomorphology of India that have certainly ameliorated our understanding of the forms and processes on the Indian coast.

Last few years have seen significant advances in this direction. There has been a remarkable shift from general, qualitative description of coastal processes and forms to the specific, quantitative and analytical studies of a variety of shore and near shore processes and coastal landforms ranging from micro to macro features on the Indian coast.

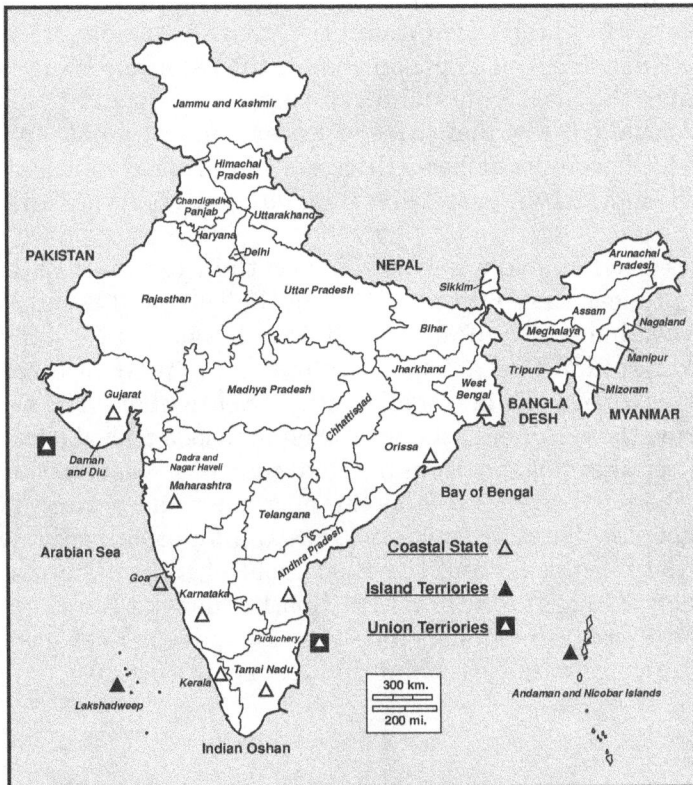

Fig. 1 : Coastal States of India

A necessity to understand the processes operating along the Indian coasts has been increasingly felt by the researchers for the proper management of the coast. This has warranted an in-depth study of sections of our coast by field visits, field mapping, field observations and measurements and interpretation of aerial photographs and satellite images of the coast. A need to date the coastal sediments is also being felt to reconstruct the geological history and palaeoenvironment of the coast.

Most of the research articles and research papers published in various books and journals discuss the landforms and analyze the processes along certain segments of the Indian coast or at certain specific sites. These placed together no doubt give an integrated picture of what the Indian coast is like geomorphologically. Majority of the accounts and research papers are based on the meticulous field measurements of various shore processes including waves, tides, littoral currents and the coastal landforms. The studies brimming with huge field data supplemented with analysis using modern techniques like raster and vector based GIS, multidimensional modelling and cell based simulations are surely giving a better and comprehensive picture of Indian coast.

The coastline of India has been undergoing morphological changes throughout the geological past. The sea level fluctuated during the period of last 6,000 years and recorded marked regression during the period between 5,000 and 3,000 years before present (Nayak,2005). The present coastal geomorphology of India has evolved largely in the background of the post-glacial transgression over the preexisting topography of the coast and offshore (Baba and Thomas, 1999).

The major rivers, the Ganga, Brahmaputra, Krishna, Godavari, and Kaveri on the east coast, and the Narmada and Tapti on the northwest coast bring large quantities of water and sediment to the coast from Indian sub continent. In addition about 100 smaller rivers, also supply considerable quantities of water and sediments to the Indian Coast. While larger rivers have well-developed deltas and estuarine systems, almost all the small rivers have estuarine mouths with extensive mud flats, salt marshes and estuarine islands (Nayak,2005).

The continental shelf of India is very wide on the west coast with about 340 km in the north, tapering to less than 60 km in the south. The shelf is narrow along the east coast and shows variable width of 35 km off Tamil Nadu to 60 km off north Andhra Pradesh and 120 km around Digha. The coastline on the west receives southerly winds that bring high waves during the monsoons (June—September). The east coast generally becomes active during the cyclones of the northeast monsoon period (October—November). The tidal range varies significantly from north to south on West as well as east Coast . It is around 11 m at the northwest, 4.5 m at the northeast and around 1 m at the south.

Considering geomorphic characteristics, the Indian coast is divided into two major divisions, namely the West coast and the East coast of India. There are presently nine coastal states namely, Gujarat,Maharashtra,Goa,Karnataka,Kerala on West coast and Tamil Nadu, Andhra Pradesh,Odisha and West Bengal on East coast. There are two coastal union territories Daman & Diu and Puducherry (Table 1).

Table1 : Coastal States of India

Coastal data

Length of coastline	7516.6 km Mainland: 5422.6 km Island Territories: 2094 km
Total Land Area	3,287,263 km²
Area of continental shelf	372,424 km²
Territorial sea (up to 12 nautical miles)	193,834 km²
Exclusive Economic Zone	2.02 x106 million km²

Maritime States and UT

Number of coastal States and Union Territories	Nine states 1. Gujarat 2. Maharashtra 3. Goa 4. Karnataka 5. Kerala 6. Tamil Nadu 7. Andhra Pradesh 8. Odisha 9. West Bengal Two Union Territories 1. Daman & Diu 2. Puducherry
Island Territories	1. Andaman & Nicobar Islands (Bay of Bengal) 2. Lakshadeweep Islands (Arabian Sea)

Coastal Geomorphology (Mainland)

Sandy Beach	43 %
Rocky Coast	11%
Muddy Flats	36%
Marshy Coast	10%
Coastline affected by erosion	1624.435 km mainland 132 km (islands)

Coastal Ecosystems

Coastal wetland area	43230 km²
Major estuaries	97
Major Lagoons	34
Mangrove Areas	31
Area under mangroves	6740 km² (57% East coast,23% west coast, 20% Andaman &Nicobar Islands)
Coral Reef Areas	5
Marine Protected Areas (MPA)	31
Area Covered by MPA	6271.2 km²

Coastal Biodiversity	
Marine Algae	217 genera 844 species
Sea grasses	6 genera 14 species
Mangroves	25 families, 43 genera, 39 species, Associated flora: 420, Associated fauna: 1862
Crustaceans	2934 species
Molluscs	3370 speceis
Echinoderms	765 species
Hard Corals	218 species
Fishes	2546 species
Reptiles	5 sea turtle species, 26 sea snake species
Marine Mammals	25 reported from Indian waters, 3 species of cetaceans: Irrawaddy dolphin, Ganga River Dolphin and Sperm whale; Dugong listed in Schedule I of Wildlife Act 1972
Coastal Length of Indian States	WEST COAST • Gujarat — 1600 Km , • Maharashtra — 720 Km, • Goa — 101 Km, • Karnataka — 320 Km , • Kerala — 580 Km EAST COAST• Tamil Nadu — 1076 Km, • Andhra Pradesh — 974 Km, • Odisha — 480 Km , West Bengal --- 220 km.

Source : Centre for coastal zone management and coastal shatter belt Updated On : 17/03/2016

Coastal Evolution

The west coast is backed by the Sahyadri virtually all along its length. Rifting, separation from Madagascar and drifting since Late Cretaceous (Table 9) has led to the formation of the west coast sedimentary basins (Kumaran *et al* ,2012). The west coast sedimentary basins exhibit patchy Holocene sediments confined to estuaries and creeks. According to kumaran *et al* (2012) the occurrence of Neogene (23 MYBP) and Recent deposits as longitudinal outcrops paralleling the western coast has led to the view that Kerala coast was under the influence of local marine environments to a minor extent in the geological past.

The conceptual models of the evolution of various land forms along the Indian coastline have been prepared recently on the basis of stratigraphic record along with geochronological data. These are surely helpful in knowing the evolution of Indian coasts.

It is now confirmed that the eastern coastline of India originated in the post-Cretaceous times (Table 9) though it was modified considerably during the Quaternary due to progradation of the deltas and impacts of glaciations and deglaciation. The evolution of east coast had a profound effect in shaping the east coast basins (Kumaran *et al* ,2012)

The coastline comprises of headlands, promontories, rocky shores, sandy spits, barrier beaches, open beaches, embayment, estuaries, inlets, bays, marshy land and offshore islands.

These coastal and shoreline environments have been strongly affected by sea level changes throughout the geological history and are likely to be affected in the near future (Hashimi,1992). The Indian mainland comprises of nearly 43% sandy beaches, 11% rocky coast with cliffs and 46% mud flats and marshy coast (Table 2, V.Sanil Kumar *et al*,2006).

Table 2: Types of coastline in different maritime states of India

State	Sandy beach(%)	Rocky coast (%)	Muddy flats(%)	Marshy coast(%)	Total length of coast* (km)	Length affected by erosion** (km)
Gujarat	28	21	29	22	1214.7	36.4
Maharashtra	17	37	46	—	652.6	263.0
Goa	44	21	35	—	151.0	10.5
Karnataka	75	11	14	—	280.0	249.6
Kerala	80	5	15	—	569.7	480.0
Tamil Nadu	57	5	38	—	906.9	36.2
Andhra Pradesh	38	3	52	7	973.7	.2
Orissa	57	—	33	10	476.4	107.6
West Bengal	—	—	51	49	157.5	49.0
Daman and Diu					9.5	—
Pondicherry					30.6	6.4
Total mainland	43	11	36	10	5422.6	1247.9
Lakshadweep					132.0	132.0
Andaman and Nicobar					1962.0	
Total					7516.6	1379.9

Source : (V.Sanil Kumar et al, 2006)*According to the Naval Hydrographic Office. **Information collected from respective states.

The West coast of India is different from the East coast in many respects. Absence of major deltas is a characteristic of west coast. Headlands, promontories, bays, creeks and estuaries, lagoons characterize the West coast. There is distinct evidence of the effect of neotectonics in some sections (Vaidyanathan, 1987). It has many problems related to tectonism and regression and transgression records of fluctuating sea level. The east coast on the contrary is known for the number of deltas especially along the northern portion, West Bengal and Odisha coast. Delta areas in the southern portion along east coast distinctly show some ancient channels, ancient beach ridges, former confluences, and strandlines.

Recent research contributions on coastal geomorphology of India show a serious concern about the degradation of environment on East and West coast. The current research in India is mainly aimed at suggesting measures and overall management of coastal environment. Accelerated erosion, progressive siltation and the overall degradation of marine eco-systems are the main problems of Indian coasts and they are being addressed by many scholars.

Ample literature is now available on Geomorphology, Geography and Geology of various sectors of Indian Coasts. There is a clear emphasis on new techniques and methodology for the understanding of processes and landforms on this coast.

Most of the early literature on Indian coastal geomorphology was essentially of a descriptive nature based on the nature, location, and relationships of the landforms and sea level. Ahmad's (1972) was possibly the first and only book on coastal geomorphology of India. It was however based on large-scale maps. In addition, there were some isolated studies by Vaidyanathan (1987), Baba and Thomas (1999). The Space Application Centre (SAC, 1992) has carried out a comprehensive study on the coast using LANDSAT and IRS data. Sea level variation and its impact on coastal environment (1990) by G.Victor Rajamanickam, Quaternary deltas of India (1991) by R. Vaidyanadhan, Coastal Geomorphology of Konkan (1993)and coastal processes and landforms(2009) by Shrikant Karlekar, Quaternary sea level variation, Shore line displacement and Coastal environment by G.Victor Rajamanickam and Michael Tooley (2001), and Coastal geomorphology and environment(2002) by A.K. Paul need a special mention as they are devoted exclusively to specific aspects of coastal geomorphology and are outstanding contributions to the coastal Geomrophology of India.

The reference list provided in this book amply demonstrates the above aspects of coastal geomorphology in India. A good number of articles referred are worth mentioning for their original contribution in the understanding of Indian coasts. These articles are devoted to one or the other aspect of coastal geomorphology and are published in the journals and books related to environment, geology, geography and general geomorphology. Mention must be made of the research journals which frequently publish research articles and papers on various aspects and areas of coastal geomorphology in India. They are, Indian Journal of Geomorphology (Allahabad), Journal of The Indian Association of Sedimentologist (Aligarh), Journal of Geological Society of India (Bangalore). Geographical Review of India (Kolkata), Transactions, Institute of Indian Geographers (Pune), Indian Journal of Marine Sciences (New Delhi), Indian Journal of Geo-Marine Sciences (New Delhi) and journal of coastal research (Online journal).

Chapter 1
THE CONTINENTAL SHELF OF INDIA

The west coast of India, extending from Sir Creek in the north (23°58' N, 68°48' E) to Kanyakumari (8°22' N 77°34' E) in the south, is a trailing passive margin that bears the imprint of generally shored parallel structural elements(Faruque et al,2014). The width of continental shelf varies from 345 km off Daman in the north to 120 km off Goa and tapers to 60 km off Kochi in the south (Fig. 2). The western continental shelf of India has an area of about 310 000 sq.km. The mid-shelf is uneven topographically, and the outer shelf is interrupted by shore-parallel ridges and reefs with a relief of 2–18 m. It is divided into four sedimentary basins, namely Kutchchh, Mumbai, Konkan and Kerala (Ramaswamy,1980).

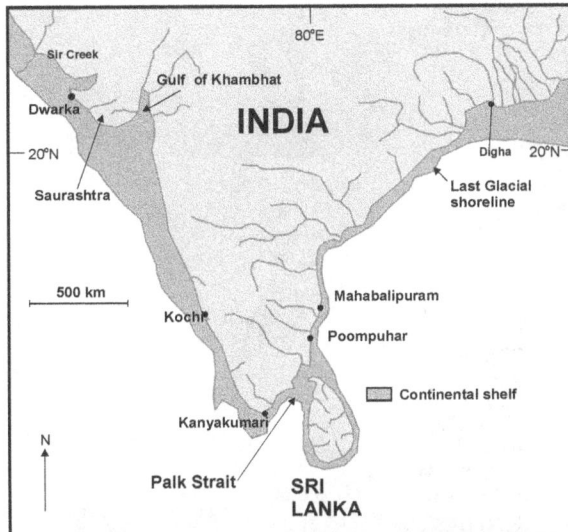

Fig. 2 : Continental shelf of India

The 2493 km-long shoreline of eastern India is fringed with a continental shelf with a variable width of 35 km off Tamil Nadu to 60 km off north Andhra Pradesh and 120 km around Digha (Fig.2). The shelf has a gentle slope in the northern sector and is moderately steep in the south. The eastern continental margin of India is aligned NE–SW due to the trend of the Eastern Ghat orographic belt in the northern part and almost north–south in the southern part. It is characterized by four major deltas; the Ganga, Mahanadi, Krishna–Godavari and Kavery.

Sediments on Continental shelf :

Primary controls on continental shelf sedimentation are tectonic and climatic (Hashimi, 1992).The sea level also exerts a strong influence on the nature and amount of sediment deposited on ocean floor.

According to Hashimi (1992) the western shelf of India is floored with three different types of sediments. The near shore sand zone extending from the shoreline to a water depth of 5-10 m, followed by the muds (silt and clay) which extend to a water depth of 50-60 m. (Fig. 3) The shelf beyond 50-60 m depth is covered by coarse calcareous sands. According to Guptha(1979) the West coast continental shelf is divided into an inner shelf with modern clayey silt and silt clay sediments with high organic matter and low carbonate content, and an outer shelf having relict carbonate sediments, coarse sands with low organic matter and high carbonates. The abundance peaks of fragmented shells correspond to the terrace depths of -65, -75, -85 and -92 m in the outer shelf region off Ratnagiri which indicate the different phases of eustatic sea levels. It is inferred from the occurrence of abundance peaks at similar depths in the other regions of the continental shelf that they must correspond to the presence of lowered strands.

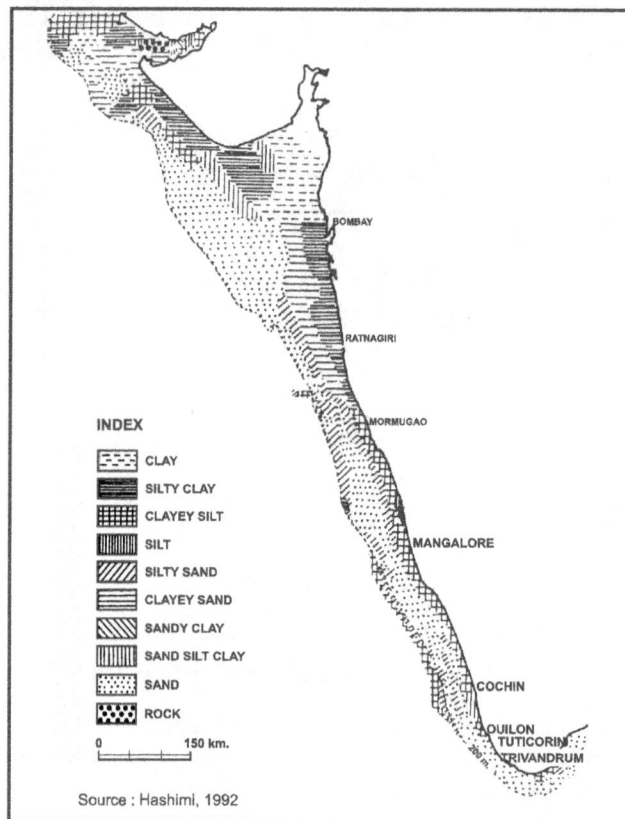

INDEX

CLAY
SILTY CLAY
CLAYEY SILT
SILT
SILTY SAND
CLAYEY SAND
SANDY CLAY
SAND SILT CLAY
SAND
ROCK

0   150 km.

Source : Hashimi, 1992

Source : Hashimi 1992

Fig. 3 : Sediments on Western Continental Shelf

2 / Coastal Geomorphology of India

The calcareous sands of the outer shelf are of Late Pleistocene origin when the sea level was 60 to 90 m below the present level. The radio-carbon dates of the calcareous sediment between 9,000 and 11,000 years before present suggest that the sediments are relict in nature. Presence of number of submerged terraces (Hashimi and Nair, 1976) at the depths of 92, 84, 71, 65, 55, 31 metres is a clear evidence of lower sea levels in the area. The most prominent of them is 90 m terrace known as Fifty Fathom Flat (Fig. 4).

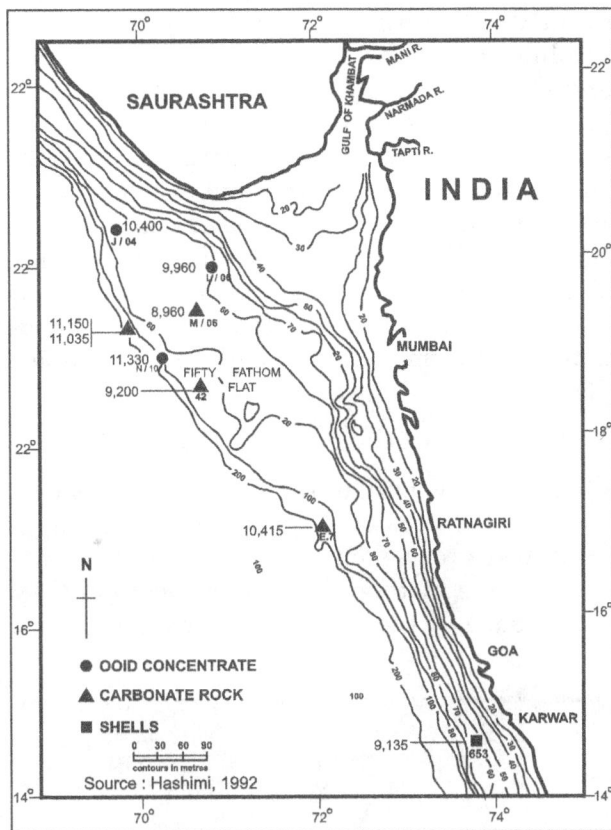

Fig. 4 : Radio Carbon Dates for Western Continental Shelf

Hashimi (1992) is of the opinion that the texture of inner shelf sediments (10 to 60 m water depth) is determined by the size and number of estuaries on the coast. Off-shore from regions where there are a large number of estuaries, the inner shelf sediments are fine grained (average mean size 5.02 Phi), rich in organic matter (> 2%) and low in calcium carbonate (< 25%). In contrast, in regions with relatively fewer estuaries, the shelf sediments are of coarser size (average mean size 1.53 phi), poor in organic matter (< 1%) and rich in calcium carbonate (> 30%). These differences are attributed to the fact that the estuaries act as regional filters which permit deposition of only fined grained sediments on the inner shelf while trapping the coarse grained material in the estuarine basin.

The interdeltaic shelves on East coast are generally sediment starved. The inner shelf is silty to clayey silt and sandy, whereas the outer shelf has carbonate sands with coral debris and shell fragments. The outer shelf is characterized by carbonate sands, lime mud and ooids in the north Andhra Pradesh sector off the Kakinada–Kalingapatnam sector. In Tamil Nadu, the region between 10°N and 12°30' N is characterized by two mega lineaments and associated tectonics, suggesting that the area is tectonically active (Faruque et al,2014).

The ^{14}C dating of relict corals, from water depths of 120 m, off Karaikal indicates an age of 18 390±210 years BP, establishing the low sea-level position of the Last Glacial Maximum. The large volume of sediment input from the deltaic system of two major drainages, Krishna and Godavari, has a significant influence on the morphology and sedimentation of the continental shelf. The influence of glacio-eustasy is noticeable in outer shelf sediments along almost the entire length of the shelf (Faruque et al,2014).

Near Puri and Kakinada a zone of lime deposition has been delineated amid terrigenous mud of the continental slope and the coastal sands and clays. This zone pinches out under the silty clays of the Godavari River to the south and those of the Mahanadi River to the north. Lime deposition has taken place mostly in the form of oolites and foraminifera. The oolites are found to be non-indigenous to the depths where they are found now. They were probably formed at a time of lowered sea level, during Pleistocene glaciations (M.Subba Rao 1964)

Delimitation of the outer limits of the continental shelf :

The continental shelf as defined under the United Nations Convention on Law of the Sea comprises of the seabed and subsoil of the submarine areas of a coastal State that extend beyond its territorial sea to the outer edge of the continental margin (comprising the shelf, slope and rise), or to a distance of 200 nautical miles from the territorial sea baselines where the outer edge of the continental margin does not extend up to that distance (Fig. 5).

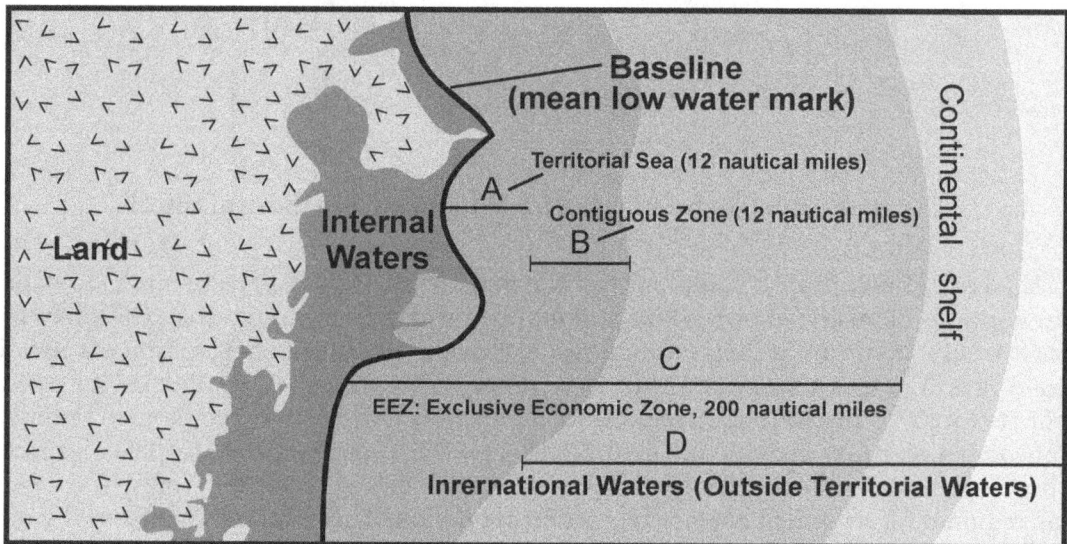

Fig. 5 : Territorial Sea

Considering that India's continental shelf extends beyond the 200 nautical miles from the territorial sea baselines, the Ministry of Earth Sciences (MoES) undertook a major multi-institutional national programme of collecting, processing, analyzing and documenting the requisite scientific and technical information for delineating the outer limits of the continental shelf in the Arabian Sea and the Bay of Bengal including the western offshore areas of the Andaman-Nicobar Islands between 2002 and 2004.

In one of the largest ever marine geophysical surveys conducted by India, over 31,000 line km of multichannel seismic reflection, gravity and magnetic data together with bathymetric information were acquired. From this information and the knowledge of the depth to the seabed, the thickness of the sediments above the basement was calculated. The magnetic and gravity data were utilised to gather additional evidence on the nature of the basement. The data gathered were interpreted in next two years.

The studies carried out by the Indian scientists have provided a wealth of scientific data of the nature of the seabed and sub seabed in and off the Indian EEZ. The data interpretation has been a challenging exercise mainly due to two of the world's thickest accumulations of sediments on the seabed (the "Indus Fan" in the Arabian Sea and the "Bengal Fan" in the Bay of Bengal) derived from the Himalayas and the western offshore of Andaman-Nicobar islands. Further detailed studies on the data collected are expected to provide the scientific community with answers such long-debated questions as the history and geological evolution of the seas and offshore sedimentary basins around us, the origin and evolution of such enigmatic features such as the 85° East Ridge in the Bay of Bengal, the Laxmi and Laccadive Ridges in the Arabian Sea, the Gulf of Mannar, the offshore extent of the Deccan volcanics, the reasons for the association of gravity lows in the Bay of Bengal with structural highs, the development of the fans vis-a-vis the origin and growth of the Himalayas etc.

The studies also open a new vista in the exploration for hydrocarbons in the offshore areas of the extended continental shelf beyond 200M. In addition, the data gathered is expected to provide specific insights related to such areas as marine ecosystems, unconventional energy, mineral resources, and hazards resulting from extreme events, such as earthquakes and tsunamis.

India already has 12 nautical miles of territorial sea and 200 nautical miles of the Exclusive Economic Zone (EEZ) measured from the baselines. With the anticipated addition of approximately 1.2 million square km of extended continental shelf India's seabed-sub seabed area would become almost equal its land area of 3.274 million sq. km.(Ref: National centre for Antarctic and Ocean research, NCAOR, Ministry of Earth sciences, Govt.of India).

Chapter 2
COASTAL PROCESSES ON INDIAN COASTS

The geographic, geologic, meteorological and oceanographic environmental factors along the Indian coast vary significantly from one sector of the coast to the other. The sediment supply and all other coastal processes such as waves, currents and, tides are influenced by these variations. Understanding Geomorphology of Indian coasts requires a thorough knowledge of all these processes acting along the coast. Information on winds, waves, tides, currents and rate of sediment transport along a coast is required for planning and design of coastal facilities.

The Indian coast is dominated by three seasons, viz. southwest monsoon (June to September), northeast monsoon (October to January) and fair weather period (February to May).

Sediment supply :

The quantum of sediment supplied to the littoral plays a very important role in the process of siltation and erosion of the coast. The primary source of the sediments deposited on the coasts is the weathering of land. These sediments are transported through rivers to the ocean. The contribution of shelf erosion to suspended sediments in the ocean is unknown and appears to be of a very low order. The quantities of sediment contributed by headland erosion and aeolian transport are both less than 2 per cent of river transport. Another main source of sand for a particular region can be of an eroding up coast cliff and/or beach. Rivers are the major sources of sediment for the beach deposits on the Indian coast (P. Chandramohan *et al*, 2001).

The single largest source of sediment for the Arabian Sea is the river Indus which discharges about 0.45×10^{12} kg of sediment annually (Guptha and Hashimi 1985), resulting over a course of time in the formation of Indus cone with as much as 2500 m of unconsolidated sediment at its proximal end. There are 14 major rivers, 44 medium rivers and more than 200 minor rivers along the Indian coast, which are acting as predominant sources for the littoral drift. The annual discharge of sediments through these rivers into the sea is about 1.2×10^{12} kg, which accounts roughly 10 per cent of the global sediment flux to the world ocean.

Next to rivers, the headlands and beach erosion contribute significantly as sources along the Indian coast. In addition to this, direct runoff and rainfall contribute to the loss of sediments as rain-wash from sub-aerial portion of the beach. Another minor loss is due to the mining of beaches for sand and placer deposits (P. Chandramohan et al, 2001).

The sediments carried by the rivers and by the surf zone currents as littoral drift get partly deposited along the Indian coast. The Gulf of Kachchh, Gulf of Khambhat, Gulf of Mannar, Palk Bay and Sand heads act as major sinks of these sediments. Deposits in the gulf, tidal marshes, bays, beach deposits and aeolian inland transports are found to be the primary sinks for sediments moving along the Indian coast.

Estimated longshore sediment transport rates on West and East coast (Table3) show that the net transport along the east coast of India is mainly towards the north, whereas along the west coast it is mostly towards the south (V.Sanil Kumar, 2006).

Table 3 : Sediment Transport Rate at Different Locations

Location	Net transport (m3/yr)		Gross transport (m3/yr)
West coast of India			
Kalbadevi	118,580	South	147,621
Ambolgad	189,594	South	299,997
Vengurla	53,040	South	120,141
Calangute	90,000	South	120,000
Colva	160,000	North	160,000
Arge	69,350	North	200,773
Gangavali	142,018	South	177,239
Kasarkod	40,186	North	77,502
Maravanthe	25,372	North	29^36
Malpe	14,169	South	106,641
Padubidri	89,358	South	385,469
Ullal	36,165	South	38,273
Kasargod	736,772	South	958,478
Kamm	19,434	South	561,576
Kozicode	114,665	South	256,697
Natika	192,818	North	660,276
Andhakaranazhi	202,096	South	599,484
Alleppey	16,929	North	62,519
Kollam	383,784	South	805,296
Thinivananehapuram	99,159	North	1231,153
Kolachel	302,403	West	943,500

Location	Net Transport (m3/yr)		Gross Tansport (m3/yr)
East coast of India			
Ovari	1,500	South	251,300
Tiruchendur	64,100	North	87,500
Kannirajapuram	117,447	North	145,979
Naripayur	36,600	South	122,500
Muthuipettai	5,200	South	8,900
Pudhuvalasai	5,300	South	42,900
Vedaranivam	51,100	North	94,100
Nagore	96,000	South	433,000
Tarangampadi	200,600	North	369,400
Poompuhar	146,000	North	478,800
Pondichery	134,400	North	237,000
Periyakalapet	486,900	North	657,600
Tilckavanniplem	177,000	North	405,000
Gopalpur	830,046	North	94?,520
Prayagi	887,528	North	997,594
Puri	735,436	North	926,637

Source : V.Sanil Kumar, 2006

Winds :

The average wind speed during the southwest monsoon period on Indian coast is about 35 km/h (9.7 m/s), frequently rising up to 45–55 km/h (12.5–15.3 m/s). The average wind speed during northeast monsoon prevails around 20 km/h (5.6 m/s). During cyclonic period, wind speed often exceeds 100 km/h (27.8 m/s). Tropical storms known as cyclones frequently occur in the Bay of Bengal during October–January.

Basic wind speed at 10 m above ground level along the Indian coast varies from 39 to 50 m/s. (Table 4).

Waves :

The maximum significant wave height recorded during the passage of a cyclone along the west coast of India in a water depth of 27 m is about 6 m. The west coast of India experiences high wave activity during the southwest monsoon with relatively calm sea conditions prevailing during rest of the year. On the east coast, wave activity is significant both during southwest and northeast monsoons. Extreme wave conditions

Table 4 : Basic wind speed at 10 m height for some locations

Location	Lat/Long (Deg. Dec.)	Basic wind speed(m/s)
Mumbai	18.9/72.8	44
Panaji	15.5/73.8	39
Mangalore	12.8/74.8	39
Kozhikode	11.2/75.8	39
Port Blair	83 /73.0	39
Thiruvananthapuram	8.5/765	39
Puducherry	11.9/79.8	50
Chennai	11.9/79.8	50
Port Blair	11.6/92.8	44
Visakhapatnam	17.7/83.3	50
Kolkata	22.5/88.3	50

Source : (V.Sanil Kumar, 2006)

occur under severe tropical cyclones, which are frequent in the Bay of Bengal during the northeast monsoon period. Along the west coast, waves approach from west and WSW during southwest monsoon, west and WNW during northeast monsoon and southwest during fair weather period. On the east coast, waves approach from southeast during south-west monsoon and fair weather period and from northeast during northeast monsoon.

Significant wave heights is for 100 year return period and average wave period reported by various agencies is given in Table 5 below (V. Sanil Kumar,2006).

Longshore Currents :

The longshore sediment transport rate reported for different locations along the Indian coast shows local reversals in the transport direction in a number of locations along the west coast. The sediment deposition/siltation noticed at most of the harbour channels and river mouths are mainly due to their interference to free passage of longshore sediment transport. Along the east coast, longshore transport is southerly from November to February, northerly from April to September and variable in March and October. Along the west coast, longshore sediment transport is generally towards the south from January to May and in October (Table 6).

The coasts near Malvan, Dabhol, Murud and Tarapur on west coast appear to be nodal drift points with equal volume of longshore sediment transport in either direction annually. Coasts near Tarangampadi, Karaikal, Nagore, Tuticorin, Virapandiapattinam and Manakkodam in Tamil Nadu behave like nodal drift points, with an equal volume of

Table 5 : Wave Chararteristics at Different Location Based on Measured Data

Location	Lat/Long	Water Depth (m)	Duration of Data (month)	Hs for 100 Year Return Period (m)	Predominant Average Wave Period (s)
Mundra*	22.8/69.7	18	6	4.4	2-6
Kandla*	22.9/70.2	15	12	4.4	3-11
Dahej*	21.7/72.6	20	7	3.0	2-10
Hazira*	21.1/72.6	15	7	4.2	2-14
Daman*	20.4/72.8	27	7	8. OT	3-15
Umbergaon*	20.2/72.7	37	6	6.6T	3-16
Vadhavan point *	19.9/72.6	24	8	3.4	3-15
Bombay high*	19.4/71.3	75	19	7.8	3-16
Uran*	18.9/72.9	10	5	3.2	4-16
Dabhol*	17.6/73.2	14	5	7.1	3-8
Jaitapur*	16.6/73.4	16	12	5.6	3-15
Mormugao*	15.4/73.8	23	12	931	3-9
Kalpeni*	10.1/73.6	11	12	3.1	3-12
Androth*	10.8/73.7	11	12	3.6	3-9
Agatti*	10.9/72.2	15	12	2.2	3-11
Minicoy	8.3/73.1	10	12	2.3	3-11
Kannirajapuram *	9.1/78.4	12	12	2.3	3-9
Nagore*	10.8/79.9	10	12	3.2	2-9
Pillaiperumalnallur*	11.1/79.9	15	12	2.7	2-9
Parangipettai*	11.5/79.8	80	6	2.8	3-12
Machalipattanam*	16.2/81.2	20	8	3.0	3-15
Narasapur*	16.4/81.7	10	11	4.6	4-12
Yanam*	16.7/82.2	90	12	3.5	3-15
Tikkcavanjialem*	17.6/83.1	12	12	4.2	3-10
Gopalpur*	19.2/84.9	10	12	3.1	3-9
Paradip*	20.3/86.0	16	9	6.3	3-10
Sunderban*	22.0/89.2	20	3	2.1	NA.

*Based on data collected by NIO, Goa,

**Based on data collected by Central Water and Power Research Station, Pune.

Source : *(Based on Source: V.Sanil Kumar, 2006)*

Table 6 : Currents at shallow Water Along the Indian Coast

Station	Water Depth (m)	Location Distance from Bed(m)	Period	Speed (m/s)	Predominant Direction(deg)
Kharo creek	18	10	December 1994	0.3-1.0	240.300
Positra	20	10	December 1993	0.40.5	180.360
Kandla	10	4	March 1996	0.05-1.6	180.360
	10	4.5	October 1996	0.05-1.5	180.360
Vadinar	25	20	Marchl 994	0.2-0.8	60.270
Muldwaraka	17	11	January 2000	0.1-0.5	290. 120
Dahej	24	23	July to October 2003	0.01-3.2	180.360
Dhabol	9	5.6	October 1994	0.1-0.5	330. 150
Mormugao	23	10	April 1998	0.02-1.2	180.360
	23	7	January 1998	0.03-0.3	180.360
	5	2.5	April 1996	0.03-0.6	120-300
	8	2.5	November 1995	0.05-0.5	90-270
	3	1.5	April 1996	0.02-0.6	110-270
	3	1.5	September 1996	0.1-0.6	110-300
Karwar	10	5	May June 1988	0.02-0.5	90.270
	14	7	May June 1988	0.02-0.6	180-270
Mangalore	9	6	May 1999	0.05-0.4	180.360
Kochi	8	6	April 1998	0.05-0.4	170-220
	6.5	4.5	April 1998	0.05-0.9	170-350
	4	3	October 1998	0.05-1.4	180-270
Kanrdajapuram	4	2	February 1997	0.01-0.3	215
	4	2	August 1997	0.01-0.2	30-60
	4	2	December 1997	0.01-0.2	210
	7	3.5	March 1997	0.01-0.3	215
	7	3.5	July 1997	0.01-0.3	45-360
	12	6	Mirch 1997	0.1-0.9	270
	12	6	August 1997	0.1-0.9	60
Nagapattinam	16	8	February 1995	0.12-0.6	30.330
	16	8	August 1995	0.040.4	180.360
	14	7	March 1995	0.1-0.4	180.360
	14	7	September 1995	0.11-0.5	180.360
Chinnakuppam	14	7	August 1996	0.03-0.3	180-225
MahabalipuRam	9	7	March 1996	0.1-0.3	30-360
Tikkavaniplem	12	9	January 1998	0.02-0.3	90.270
Gopalpur	15	3	January 1994	0.1-0.4	225
	15	7	February 1994	0.05-0.4	45.225
Paradip	15	9	May 1996	0.1-0.8	65
	30	29	November 1996	0.1-1.2	30-60

Source : (V.Sanil Kumar, 2006)

transport in either direction annually. Annual net transport is northerly on the east Gujarat coast. The coast between Pondicherry (Puduchery) and Point Calimere in Tamil Nadu, and Maharashtra coast experiences negligible quantity of annual net transport. The annual net transport at the southernmost tip of the Indian Peninsula (Kanyakumari) is almost negligible. The annual net sediment transport at Visakhapatnam is northerly, which is similar to that observed at Tikkavanipalem, 30 km south of Visakhapatnam. (V. Sanil Kumar, 2006).

The maximum currents vary from about 1.4 m/s in the open coast to about 3.2 m/s in the Gulf of Khambhat. The gross longshore sediment transport rate is about $1x10^6$ m³/yr along south Kerala and south Orissa. It has been recorded that the net transport along the east coast of India is towards the north, whereas along the west coast it is mostly towards the south.

The Tides :

Tides are the rise and fall of sea levels caused by the combined effects of the gravitational forces exerted by the Moon and the Sun and the rotation of the Earth. The tidal range is the vertical difference between the high tide and the succeeding low tide. The most extreme tidal range occurs around the time of the full or new moons, when the gravitational forces of both the Sun and Moon are in phase, reinforcing each other in the same direction (new moon), or are exactly the opposite phase (full). This type of tide is known as a spring tide. During neap tides, when the Moon and Sun's gravitational force vectors act in quadrature (making a right angle to the Earth's orbit), the difference between high and low tides is smaller. Neap tides occur during the first and last quarters of the moon's phases. The largest annual tidal range can be expected around the time of the equinox, if coincidental with a spring tide (Wikipedia) The exact tidal range at a place depends on the volume of water adjacent to the coast, coastal processes and morphology and the configuration of the coastal stretch.

Spatially, tidal range varies according to the hydrodynamic response of a particular ocean basin, shelf sea, bay or estuary to astronomical forces generating tides. Tidal ranges are usually largest in semi-enclosed seas and funnel-shaped entrances of bays and estuaries, or regions where a continental shelf has the right combination of depth and width for tidal resonance to occur. Conversely, tidal ranges are typically smallest in the open ocean, along open ocean coastlines and in almost fully enclosed seas. Tides are often classified by their mean range: macrotidal (> 4 m); mesotidal (2 to 4 m); and microtidal (< 2 m).

The mean tidal range along the Indian coastal region varies from 12.5 m at Bhavnagar, Gulf of Khambhat to 0.5 m along the peninsular tip of India. Along the Indian coast, the tides observed are a combination of both semi-diurnal and diurnal tides. The tides and tidal currents are weaker along the southern part of the Indian coast (current speeds of the order of a few tens of cm/s). The magnitude of tidal currents increases northward and it reaches very high values in certain areas like the Gulf of Kutch and the Gulf of Khambhat, where the speeds can exceed 2 m/s.

Table 7 gives the largest known tidal ranges on the Indian coasts

Table 7 : Largest known Tidal Ranges on Indian Coast (m)

Station	State	Lat (DD)	Long(DD)	Largest TR(m)
Kannyakumari	Tamilnadu	8.1	77.5	0.89
Manavalakurichi	Tamilnadu	8.1	77.3	0.9
Colachel	Tamilnadu	8.2	77.2	0.9
Minicoy	Lakshdweep	8.3	73.1	1.48
Kovalam	Kerala	8.4	76.9	0.9
Thiruananthapuram	Kerala	8.5	76.9	1.04
Tiruchendur	Tamilnadu	8.5	78.1	0.89
Kayalpattinam	Tamilnadu	8.6	78.1	0.89
Paravur	Kerala	8.8	76.7	1.1
Thoothukudi	Tamilnadu	8.8	78.2	0.93
Varakala	Kerala	8.8	76.7	1.09
Kollam	Kerala	8.9	76.6	1.12
Mandapam	Tamilnadu	9.3	79.1	0.8
Pamban	Tamilnadu	9.3	79.2	0.87
Thondi	Tamilnadu	9.7	79	0.27
Kochi	Kerala	9.9	76.2	1.12
Eloor	Kerala	10.1	76.3	1.35
Kodungallur	Kerala	10.2	76.2	1.38
Adirampattinam	Tamilnadu	10.3	79.4	0.39
Vedaranyam	Tamilnadu	10.4	79.9	1.12
Kavaratti	Lakshdweep	10.6	72.6	1.61
Nagapattinam	Tamilnadu	10.8	79.8	1.12

Station	State	Lat (DD)	Long(DD)	Largest TR(m)
Ponnani	Kerala	10.8	75.9	1.5
Karaikal	Puduchery	10.9	79.8	1.13
Tirur	Kerala	10.9	75.8	1.52
Tharangambali	Tamilnadu	11	79.9	1.11
Beypore	Kerala	11.2	75.8	1.55
Feroke	Kerala	11.2	75.8	1.55
Tirumullaivasal	Tamilnadu	11.2	79.8	1.11
Kozhikode	Kerala	11.3	75.8	1.56
Port Blair	Andaman	11.6	92.8	2.49
Bamboo Flat	Andaman	11.7	92.7	2.48
Cuddalore	Tamilnadu	11.7	79.8	1.15
Mahe	Kerala	11.7	75.5	1.65
Dharmadam	Kerala	11.8	75.5	1.65
Kannur	Kerala	11.8	75.4	1.68
Mulappilangad	Kerala	11.8	75.5	1.65
Thalassery	Kerala	11.8	75.5	1.65
Pappinisseri	Kerala	11.9	75.4	1.73
Payyannur	Kerala	12.1	75.2	1.81
Marakkanam	Tamilnadu	12.2	79.9	1.17
Kanhangad	Kerala	12.3	75.1	1.84
Nileshwar	Kerala	12.3	75.1	1.84
Kasargod	Karnataka	12.5	74.9	1.84
Mamallapuram	Tamilnadu	12.6	80.2	1.27

Station	State	Lat (DD)	Long(DD)	Largest TR(m)
Manjeshwara	Kerala	12.7	74.9	1.92
Mangaluru	Karnataka	12.8	74.8	1.85
Ullal	Karnataka	12.8	74.8	1.9
Perungudi	Tamilnadu	12.9	80.3	1.28
Gandhi nagar	Tamilnadu	13	80.3	1.35
Chennai	Tamilnadu	13.1	80.3	1.5
Mulki	Karnataka	13.1	74.8	1.94
Kattivakkam	Tamilnadu	13.2	80.3	1.35
Gangolli	Karnataka	13.7	74.7	2.13
Baindur	Karnataka	13.9	74.6	2.16
Bhatkal	Karnataka	13.9	74.5	2.18
Murdeshwar	Karnataka	14.1	74.5	2.2
Honavar	Karnataka	14.3	74.4	2.28
Kumta	Karnataka	14.4	74.4	2.3
Gokarna	Karnataka	14.5	74.3	2.33
Ankola	Karnataka	14.7	74.3	2.35
Karwar	Karnataka	14.8	74.1	2.45
Cancona	Goa	14.9	74	2.42
Cavelossim	Goa	15.2	73.9	2.43
Cuncolim	Goa	15.2	73.9	2.43
Varca	Goa	15.2	73.9	2.47
Benaulim	Goa	15.3	73.9	2.45
Chinchinim	Goa	15.3	73.9	2.47

Station	State	Lat (DD)	Long(DD)	Largest TR(m)
Colva	Goa	15.3	73.9	2.45
Gova velha	Goa	15.4	73.9	2.47
Marmgao	Goa	15.4	73.8	2.56
Sancole	Goa	15.4	73.9	2.46
Vasco	Goa	15.4	73.8	2.47
Bambolim	Goa	15.5	73.9	2.47
Calangute	Goa	15.5	73.8	2.47
Panaji	Goa	15.5	73.8	2.47
Taleigao	Gao	15.5	73.8	2.47
Vengurla	Maharashtra	15.5	73.6	2.56
Siolim	Goa	15.6	73.8	2.47
Vagator	Goa	15.6	73.7	2.47
Arambol	Goa	15.7	73.7	2.49
Malvan	Maharashtra	16.1	73.5	2.61
Kakinada	Andhra Pradesh	16.9	82.3	2.2
Dabhol	Maharashtra	17.6	73.2	3.28
Vishakhapattanam	Andhra Pradesh	17.7	83.2	2.15
Harnai	Maharashtra	17.8	73.1	3.43
Bhimunipatnam	Andhra Pradesh	17.9	83.5	1.94
Shrivardhan	Maharashtra	18	73	3.82
Mhasala	Maharashtra	18.1	73.1	3.7
Murud	Maharashtra	18.3	72.9	4.24
Revdanda	Maharashtra	18.6	72.9	4.74

Station	State	Lat (DD)	Long(DD)	Largest TR(m)
Alibag	Maharashtra	18.7	72.9	4.96
Mumbai	Maharashtra	18.9	72.8	5.22
Panvel	Maharashtra	18.9	73.1	5.63
Sompeta	Andhra Pradesh	18.9	84.6	2.12
Uran	Maharashtra	18.9	72.9	5.77
Navi Mumbai	Maharashtra	19	73	5.85
Airoli	Maharashtra	19.2	72.9	5.5
Borivali	Maharashtra	19.2	72.8	5.3
Thane	Maharashtra	19.2	72.9	5.44
Chatrapur	Odisha	19.3	85	2.19
Virar	Maharashtra	19.5	72.8	4.98
Ganjam	Odisha	19.6	84.7	2.2
Shirgaon	Maharashtra	19.7	72.7	4.6
Konark	Odisha	19.8	86.1	2.29
Puri	Odisha	19.8	85.8	2.27
Chinchani	Maharashtra	19.9	72.7	4.15
Dahanu	Maharashtra	19.9	72.7	4.61
Paradwip	Odisha	20.3	86	2.6
Chandbali	Odisha	20.7	86.8	3.6
Diu	Gujarat	20.7	70.9	2.82
Delvada	Gujarat	20.8	71	3.02
Kodinar	Gujarat	20.8	70.7	2.71

Station	State	Lat (DD)	Long(DD)	Largest TR(m)
Khambat	Gujarat	20.9	71.6	5.6
Pipav Bandar	Gujarat	20.9	71.5	4.38
Veraval	Gujarat	20.9	70.3	2.63
Katpur	Gujarat	21.1	71.9	5.75
Hansot	Gujarat	21.6	72.8	5.43
Sagar island	West Bengal	21.6	88	5.74
Ghogha	Gujarat	21.7	72.3	4.24
Bhavnagar	Gujarat	21.8	72.2	12.02
Dhuwaran	Gujarat	22.2	72.8	5.86
Diamond Harbour	West Bengal	22.2	88.2	6.47
Salaya	Gujarat	22.3	69.6	4.45
Sika	Gujarat	22.4	69.8	4.64
Kolkata	West Bengal	22.5	88.3	5.79
Okha	Gujarat	22.5	69.1	4.69
Dwaraka	Gujarat	22.7	68.9	3.61
Jodiya Bandar	Gujarat	22.7	70.3	5.4
Mundra	Gujarat	22.8	69.7	4.63
Kandla	Gujarat	22.9	70.2	8.54
Navalakhi	Gujarat	22.9	70.5	8.95

Source : *(Microtidal type TR < 2 m, Mesotidal type TR 2 to 4 m, Mesotial type TR > 4 m)*
Tidal gauge stations

Chapter 3

THE COARAL ISLANDS AND ISLAND TERRITORIES

Coral Islands :

Coral islands in Arabian sea and Bay of Bengal appear either as an accumulation of coral sand and gravel on the surface of coral reefs or as a slightly emerged limestone platform of formerly live coral not more than a few meters above mean low water.

(A) Gulf of Kachchh Islands :

The 42 islands of the Gulf of Kachchh (22°15'N-23°40'N; 68°20'-70°40'E) are the northernmost coralline or sandstone based islands in India. Almost uninhabited, the vegetation inland consists only of shrubs. Several of the islands have dense mangrove patches on the coast, 34 islands have fringing reefs (often called as patch reefs) confined

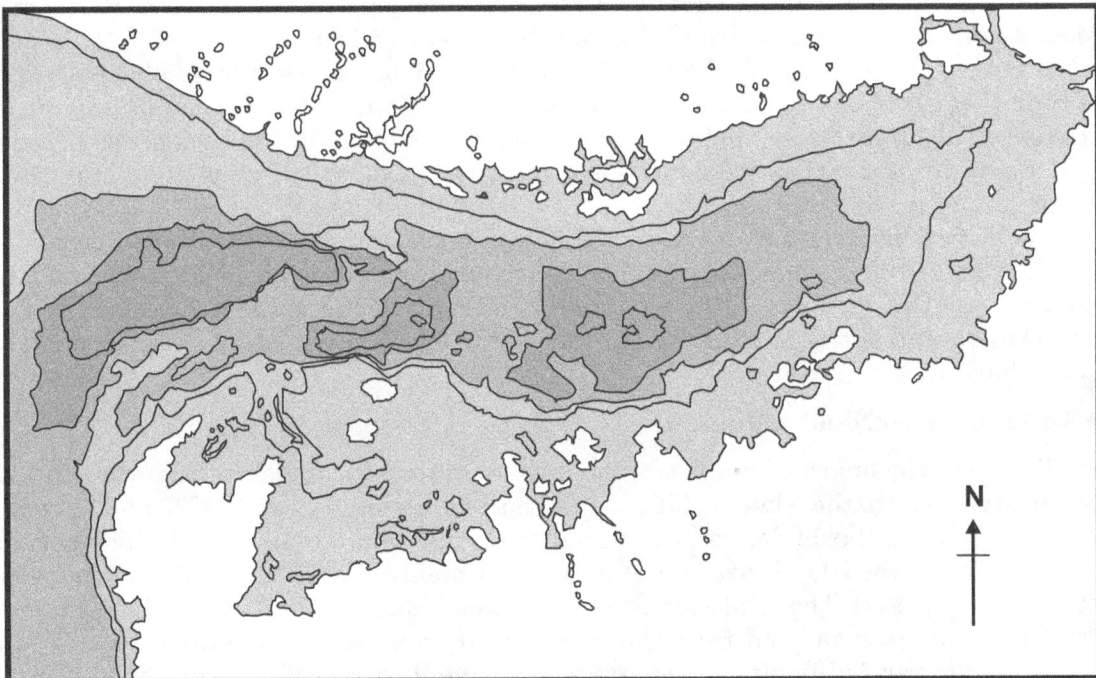

Source : CSIR / NIO

Fig. 6 : Gulf of Kuchchh Islands

to intertidal sandstones or wave-cut, eroded, shallow banks (Fig. 6). The region is tectonically unstable and evidence of uplift can be seen in the form of raised reefs near the mouth of the Gulf, not far from extant islands (Wafar *et al*, 2005).

The coastal geomorphology and the fauna and flora of the islands are influenced considerably by the sediment depositional regime, high-velocity tidal currents (up to 5 knots), and a large range in environmental parameters (e.g., temperature 15°-30°C, salinity 25-40). The extreme conditions also limit coastal biodiversity to 37 species of corals and a smaller number of other invertebrates. However, algal growth along these coasts can be substantial at certain times of the year. The mangroves already constrained by high salinity and high tidal exposure also have been heavily impacted due to felling for fuel and fodder. Areas around some of the islands have earlier been good pearl oyster and fishing grounds (Wafar *et al*, 2005).

(B) Lakshadweep Islands :

These are the northern-most islands of the Laccadive-Chagos ridge (9°-12°N; 7°-74°E). They are located about 200-400 km on the southwest coast of India. This part of the ridge comprises of 12 atolls, 3 reefs, and 5 submerged banks. Of the 36 islands on the atolls, only 10 (Minicoy, Kalpeni, Andrott, Agatti, Kavaratti, Amini, Kadamat, Chetlat, Kilian, and Bitra) are inhabited. The northernmost Bitra Island is the smallest inhabited island in India. Among these, Minicoy is separated from the rest by the 9° channel. It is culturally and linguistically closer to the Maldivian islands.

Mainly coralline and no more than a few square kilometers in area, all these islands are low lying with profuse coral growth all around. The only cultivated plant is coconut, besides a few vegetable and horticultural plants introduced from the mainland. The coast toward the lagoon is sandy and habited by sand dune flora. The sea ward shore is rocky and typical of all oceanic atolls, with a steep drop in profile. Radiocarbon dating of the storm beach at Kavaratti island gave an age of about 6000 YBP indicating their recent origin. Some of the uninhabited islands are only sand cays; one of them is an important nesting ground for seabirds.

The known biodiversity status is as follows: hard corals 104 species, soft corals-37 species, fishes—163 species, invertebrates—about 2,000, and algae-119 species. The bleaching event of 1998 has however caused a serious reduction in coral biodiversity. Shore erosion and silting are additional causes for loss of coral cover and reduction in species abundance.

(C) Andaman and Nicobar Islands :

These are the emerged parts of a mountain chain that stretches from the Arakan Yoma in Myanmar to the islands of Indonesia. Spread between 6° and 14°N, and between 91° and 94°E in the Bay of Bengal, these islands number more than 500, of which only 38 are inhabited. All the islands are mountainous, sedimentary in nature, and have fringing reefs towards the east. The total area covered by these islands is 8,293 km². The Andaman group of islands is separated from the Nicobar group by the 10° channel which has a heavy tidal flow and difficult to navigate with conventional crafts. Most parts of these islands are covered with thick forests and the low lying areas are covered with mangrove swamps.

Island Territories :

(A) Lakshadweep Islands :

Lakshadweep islands consist of a group of tiny coral islands, located in the Arabian Sea, about 400 km from the main land (southern tip of the Indian Peninsula). The Union Territory of Lakshadweep consists of 10 inhabited islands, 16 uninhabited islands, attached islets, 4 newly formed islets and 5 submerged reefs. These islands are scattered in the Arabian Sea between North Latitudes 8° 00' and 12° 13'N and east longitude 71° 00' and 74° 00'E (Fig.7). These islands are typically a chain of low islands surrounding a shallow lagoon, consisting largely of recent sediments on top of older coral limestone. The inhabited islands are Agatti, Amini, Androth, Bangaram, Bitra, Chetlat, Kadmat,, Kalpeni, Kavaratti and Minicoy. Chetlat, Kiltan and Kadmat are closely spaced and are on the northern part of the archipelago, whereas Kalpeni is on the east central part of the group and the Minicoy Island is located in the southernmost part and far away from the other islands. Androth, having an area of 4.84 sq.km. is the largest Island, whereas Bitra, with an area of 0.1 sq.km is the smallest.

Lakshadweep islands have a delicate ecosystem with very limited fresh water resources. Though the islands receive high rainfall, the lack of surface storage and the limited ground water storage capacity, where fresh water is occurring as a small lens floating over salt water, makes fresh water a precious commodity. High porosity of the aquifers allows mixing of freshwater with sea water. Following are the inhabited islands.

1. Agatti :

Agatti is the most westerly island of Lakshadweep. It lies on the eastern arc of the coral shoal and is about 6 Km in length and about 1000 meters wide at the broadest point. Coral growths and multi coloured coral fishes abound in this lagoon.

2. Androth :

Androth is the nearest island to the mainland. It is largest island in Lakshadweep.

3. Minicoy :

It is the second largest island. Minicoy is southernmost island in Lakshadweep, crescent shaped and has one of the largest lagoons.

4. Kavaratti :

It is headquarter of the UT Administration since 1964. The beautiful calm Lagoons form an ideal spot for water sports.

5. Kadmat :

It has a very large lagoon on the western side, Long sandy beaches and excellent water sport facilities are the stellar attraction here.

6. Kalpeni :

It is known for its scenic beauty and the small islets called Tilakkam and Pitti and an uninhabited island on the north called Cheriyam.

7. Amini :

Talented craftsmen living here are famous for making walking sticks with tortoise shells and coconut shells.

8. Kiltan :

The Island is only 3 kms long. The island is thick in flora and is fertile.

9. Chetlat :

Chetlat is the northern most inhabited island.

10. Bangaram :

A Beautiful and breathtaking island in the Lakshadweep, It has been ranked among the best getaways of the world.

11. Bitra :

Bitra Island is the smallest inhabited island in the territory. Bitra is the breeding ground for a number of seabirds. iyam.

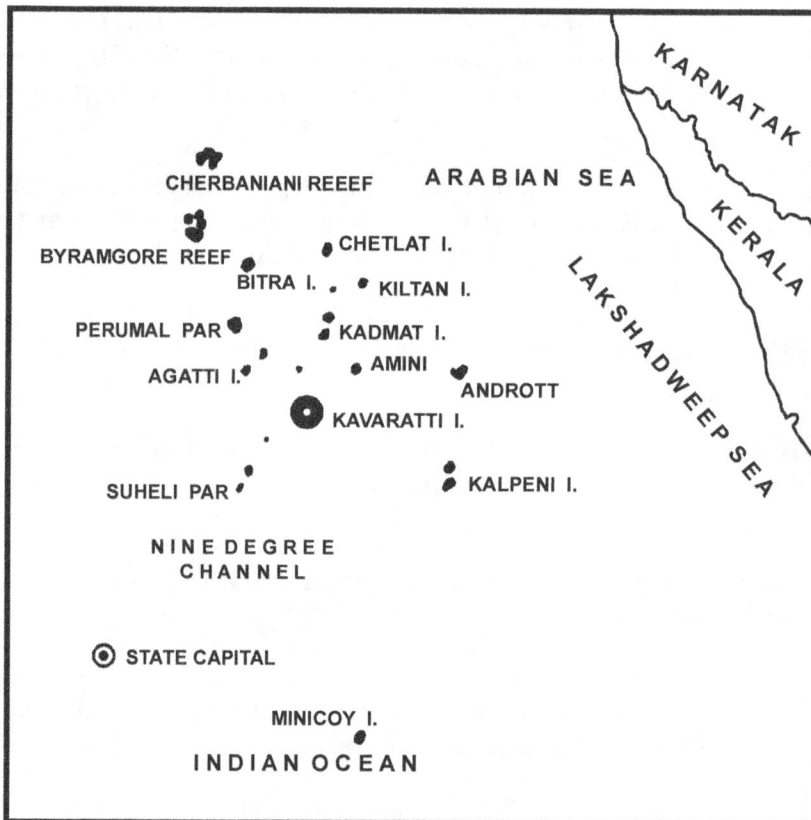

Fig. 7 : Lakshadweep Islands

Table 8 : Lakshadweep Islands: Basic Data

Latitude	8^0-12^0N
Longitude	71^0-74^0 E
Total no. of islands	36
Total no. of inhabited islands	10
Total geographical area (sq. km.)	32
Total land area (sq. km.)	26.32
Total lagoon area (sq. km.)	4200
Normal Annual Rainfall (mm)	1803
Geomorphology	
Major physiographic Units	Coral Islands –Atoll & Reef
Major Water Body	Lagoons
Major Soil types	Coral Sand
Predominant Geological Formation	Coral Limestone
Major Water bearing formation	Coral sand and Coral Limestone.

Source : (Anitha Shyam and G.Sreenath 2006: Ground Water Information Booklet of Lakshadweep Islands, Union Territory of Lakshadweep)

The total geographic area of Lakshadweep islands is 32 sq.km. The islands do not show any major topographical features but are largely low levelled and flat topped, generally rising to the height of a few metres above sea level. The height of the land above the sea level is about 1-2 m. Occasionally, old sand dunes on the sides of the lagoons and storm beaches on the seaward side of the islands rise up to height of 8 m. The storm beaches consist of coral pebbles and boulders piled up well above the high tide mark.

Most atolls have a northeast-southwest orientation with an island on the east, a broad, well-developed reef on the west and a lagoon in between. All islands of Lakshadweep are of coral origin and some of them like Minicoy, Kalpeni, Kadmat, Kiltan and Chetlat are typical atolls. The islands on these atolls are invariably situated on the eastern reef margin except Bangaram and Cheriyakara which lie in the centre of the lagoon. In the case of Bitra, the island is on the northern edge of the lagoon.

The development and growth of the islands on eastern reef margins is controlled by a number of factors. The cyclones from the east have piled up coral debris on the eastern reef while the very high waves generated annually during the southwest monsoon have pounded the reef and broken this into coarse and subsequently to fine sediments which were then transported and deposited on the eastern side behind the coral boulders and pebbles on the eastern reef (Anitha Shyam and G.Sreenath 2006)

A gradual accretion of sediments by this process has led to the growth of the islands. In some of the lagoons like Kiltan and Chetlat, the islands are growing at a very fast rate and during the next decade or so, the lagoons themselves may be filled up with sediments. In atolls where openings occur in the reef or where the lagoon is too wide for the sand to be transported across its entire width, sand banks usually develop and enlarge towards the centre of the lagoon leading to the formation of the island in the centre such as in Bangaram, Suheli etc.

The entire Lakshadweep group of islands lies on the northern edge of the 2500 km long North-South aligned submarine Lakshadweep-Chagos ridge. The Lakshadweep Sea separates this ridge from the west coast of India. The ridge rises from a depth of 2000-2700 m along the eastern side and 400 m along the Western side. The eastern flanks of this ridge appear to be steeper compared to their western counterparts. The ridge has a number of gaps, the prominent being the Nine Degree channel.

The Lakshadweep Islands are composed mainly of coral reefs and material derived from them. Barrier reefs and lagoons are seen in almost all islands. The hard coral limestone is generally exposed along the coast during low tides and is also seen in well sections. A bore hole drilled in 1972 in the 9° Channel of Lakshadweep ridge by the drill vessel 'Glomar Challenger' at a water depth of 1764 m. down to a depth of 411 m. below sea floor encountered calcareous sediments of Upper Palaeocene to Pleistocene age (Table 9). Palynological and other studies indicate that the ridge was faulted down during Lower Eocene period which resulted in the formation of Lakshadweep Sea and separation of the ridge from Peninsular India.

Coral atolls generally consist of a layer of recent (Holocene) sediments, comprising mainly coral sands and fragments or coral, on top of older limestone. An unconformity separates these two layers at typical depths of 10m to 20 m below mean sea level. Several deeper unconformities may exist due to fluctuations in sea level which results in alternate periods emergence and submergence of the atoll. During periods of emergence, solution and erosion of the reef platform can occur, while further deposition of coral limestone can occur during periods of submergence.

Lakshadweep Islands are coral islands in the Indian Waters of Arabian Sea. Several theories are available in the literature on the formation of these atolls. Darwin, as reported by Wood (1983), suggested that fringing reefs surrounding volcanic islands, during subsidence, grew up gradually to develop ring shaped barrier reefs and, as subsidence continued, eventually became centre island free ring reefs. The other theories are also equally interesting. According to Alcock (1902), all the Lakshadweep Islands are remains of eroded atolls raised only a few feet above the sea level and formed entirely of coral rock and coral sand. Gardiner (1903) believed that Maldives and Lakshdweep were formed on a large bank which was a part of an ancient land that completely sank. According to him some of the islands are remnants of mountains that existed in the sunken land. In general conformity with the geological history of the Indian Ocean reefs, it may be stated that the reefs of Lakshadweep were built in Tertiary and Quaternary eras on volcanic structures and the present day surface features of the reefs are the results of erosional and depositional consequence of Pleistocene and Holocene sea level changes (Stoddart, 1973). All the theories explain the formation of ring reefs and subsequently the atolls but not the further development of the islands.

Table 9 : Geological Time Scale

Era	System & Period	Serin & Epoch	Some Distinctive Features	Year Before Present
CENOZOIC	Quaternary	Recent Pleistocene Pliocene	Modern Man Early man; northern glaciations Large carnivores	11000 ½ to 2 million 13 + 1 million
CENOZOIC	Tertiary	Miocene Oligocene Eocene Paleocene	First abundant grazing mammals Large running mammals Many modern types of mammals First placental mammals	25 + 1 million 36 + 2 million 58 + 2 million 63 + 2 million
MESOZOIC	Cretaceous		First flowering plants; climax of dinosaurs and ammonities, followed by Cretaceous- Tertiary extinction	135 + 5 million
MESOZOIC	Jurassic		First birds, first mammals dinosaurs and ammonites abundant	181 + 5 million
MESOZOIC	Triassic		First dinosaurs, Abundant cycads and conifers	230 + 10 million
PALEOZOIC	Permian		Extinction of most kinds of marine animals, including trilobities, Southern glaciation	280 + 10 million
PALEOZOIC	Carboniferous	Pennsylvanian Mississlppian	Great coal forests, conifers First reptiles Sharks and amphibians abundant. Large and numerous scale trees and seed ferns	310 + 10 million 345 + 10 million
PALEOZOIC	Devonian		First amphibians; ammonities; fishes abundant	405 + 10 million
PALEOZOIC	Silurian		First terrestrial plants and animals	425 + 10 million
PALEOZOIC	Ordovician		First fishes; invertebrates dominant	500 + 10 million
PALEOZOIC	Cambrian		First abundant record of marine life; trilobites dominant	600 + 50 million
	Precambrian		Fossils extremely rare, consisting of primitive aquatic plants. Evidence of glaciations. Oldest dated algae, over 2600 million years; oldest dated meteorities 4500 million years	

(A) Minicoy : The island Minicoy is the southernmost in the Lakshadweep Archipelago. It lies between 8°15' to 8°20' N latitudes and 73~01' to 73"05' E longitudes with an area of 4.4 sq. Km.The island has a large lagoon on the western side measuring about 6 km across. It does not exhibit significant geomorphologic differences except for micro level relief differences. The elevation above the sea level is about 1-2 m in the west and about 2-3 m in the east during high tides. The island has been formed along the eastern fringe of the atoll covering the complete length of 12 km from southwest to northeast (Fig. 8)

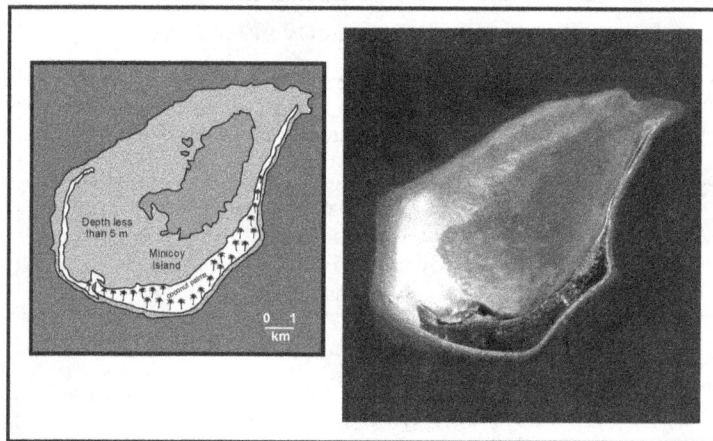

Fig. 8 : Minicoy Island

(B) Kavaratti : The island, Kavaratti lies between 10°32' · to 10°35' N latitude and 72°35' to 72°37' E longitude with an area of 3.6 sq km. The island has a large lagoon in the western side which measures about 1.5 km across. The length of the lagoon is about 6 km. It is a flat island rising above the sea level to the extent of 1 to 2 metres in general during high tides. It has a narrower southern half and a broader northern half. The western shore

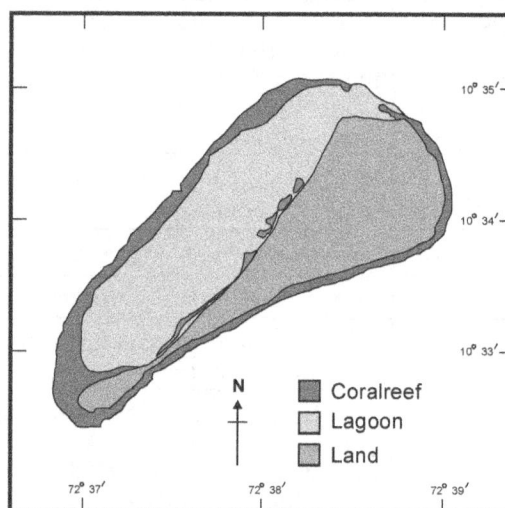

Fig. 9 : Kavaratti Island

line exhibits some undulations due to active deposition of sand. The eastern shore line has lot of beach sand stones (limestone) which act as barriers protecting the island from the cutting action of the waves (Fig. 9).

(C) **Kadmat :** The island Kadmat lies between 11⁰10' to 11⁰15' N latitudes and 72°45' to 72⁰47' E longitudes with an area of 3.1 sq. km. Kadmat is a long and narrow island. It is only 570 m wide at the broadest point. The eastern reef is exposed at low tide and forms a level platform stretching from sea beach for about 100 metres. The island has a large lagoon on the western side measuring about 2 km at the broadest point. The island is, generally, flat rising 1 to 3 m in both the east and west above the sea level during high tides. A high ridge of sand runs down the western side of the island. Also there are sand deposits in the southern part giving rise to undulations whereas the northern part is flat with little fresh sand deposits. The island is aligned north south with a slight clockwise tilt (Fig. 10).

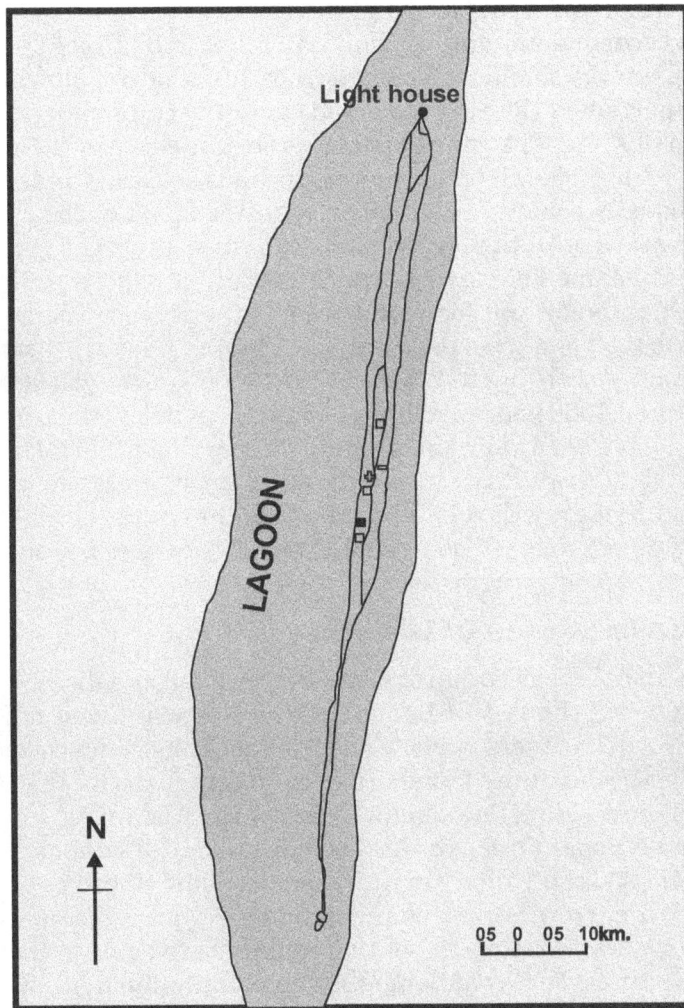

Fig. 10 : Kadmat Island

Minicoy and Kavaratti islands have formed along the eastern fringe of the atolls with a southwest to northeast orientation and Kadmat, along north-south orientation. The position and orientation of many of the Lakshadweep islands inside the ring reefs are the result of sea surface circulation and wave action which follow seasonal monsoon (southwest monsoon) winds .The sea surface circulation in the Arabian Sea is stronger and steadier during southwest monsoon compared to those in the northeast monsoon. During the southwest monsoon the surface currents in the open ocean are eastwards and clockwise in direction due to the coastal configuration. It flows north eastwards along the Arabian coast and southwards along the Indian coast as wind driven ocean current. Where reefs grow up in shallow seas, bottom currents exert a great influence on reef form right from the beginning of the reef building by depositing sediment to the leeward side of the initial coral colonies and by transporting their coral larvae which subsequently initiate reef growth on these leeward sediments (Fairbridge, 1968). This action of sea current is accompanied by the island building action of the wind and waves.

These factors become dominant as the vertical growth of reef approaches sea level. The waves gaining energy from the east and north eastward blowing monsoon winds break the western portion of the ring reefs and carry the coral shingle and sand towards the eastern portion of reefs. The ramparts of coarse shingles are then deposited at some distance in from the reef edge. The finer fragments and the coral sands are carried further on by the waves and only come to rest along the eastern part of the ring reefs as noticed in Minicoy and Kavaratii or in the lagoon when the wave energy is greatly diminished as observed in Kadmat island The islands are thus formed, whose position on the reef is determined by the frequency and force of waves and winds coming from southwest.

These islands have been growing upwards keeping pace with sea level changes in the post glacial period. According to Wood (1983) 15000 years ago the sea was about 120 m lower than today and 7000 years ago it was about 20 m below the present level. The sea level change should have been slow and gradual during this period. Had it been a sudden process these coral islands would not have existed because corals can grow in only shallow waters. The modern reef growth related to the present sea level must have begun only about less than 5000 years ago (Wood 1983). The age of these islands is less than 5000 years and their development progressed from east to west.

Andaman And Nicobar Group Of Islands :

The Andaman and Nicobar Group of Islands is situated as a dissected chain in arcuate fashion oriented in N – S the Bay of Bengal off the Eastern Coast of India and extends between 6^0 and 14^0 North Latitudes, and 92^0 and $94^°$ East Longitudes covering a geographical area of 82 sq. km. These islands are found in two major groups, with the 10° N international shipping channel as the dividing line, popularly known as Andaman Group or the Northern Group of Islands and Nicobar Group or the Southern Group of Islands. The capital town of all the groups which form the Union Territory is Port Blair (Fig. 11).

They consist of a narrow broken chain of about 572 picturesque islands, islets, and rocks extending along a general north-south direction between 14° N and 6.5° N latitude in the southeastern part of the Bay of Bengal. Of these, only about 36 islands are inhabited.

The Andaman outer arc ridge and the Barren volcano are the major tectonic features in the region. The trench along the margin in the west of the A&N islands is as deep as

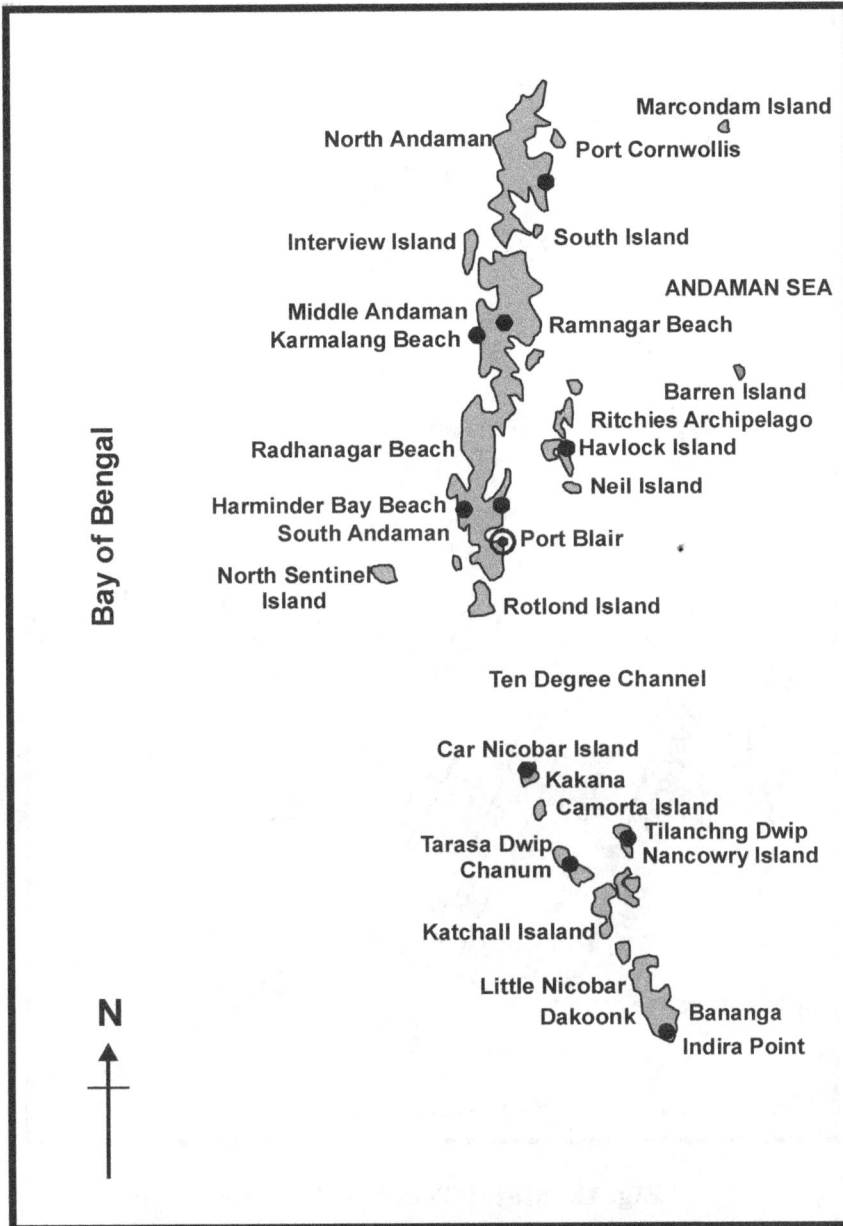

Fig. 11 : Map of Andaman & Nicobar Islands

3,000– 3,500 m (Fig 12). The ridges, indicative of a topographic high, consist of accretion of oceanic sediments, which were uplifted during the Oligocene epoch. The eastern part of the Andaman Islands is made up of highly deformed rocks from the oceanic floor of Cretaceous-Early Eocene ultrabasic/ volcanic/pelagic sediments.. The western part of the islands is occupied by sediments that belong to Eocene-Oligocene sandstone/siltstone with conglomerates along with Mio-Pliocene calcareous sediments. (Malik et al, 2006).

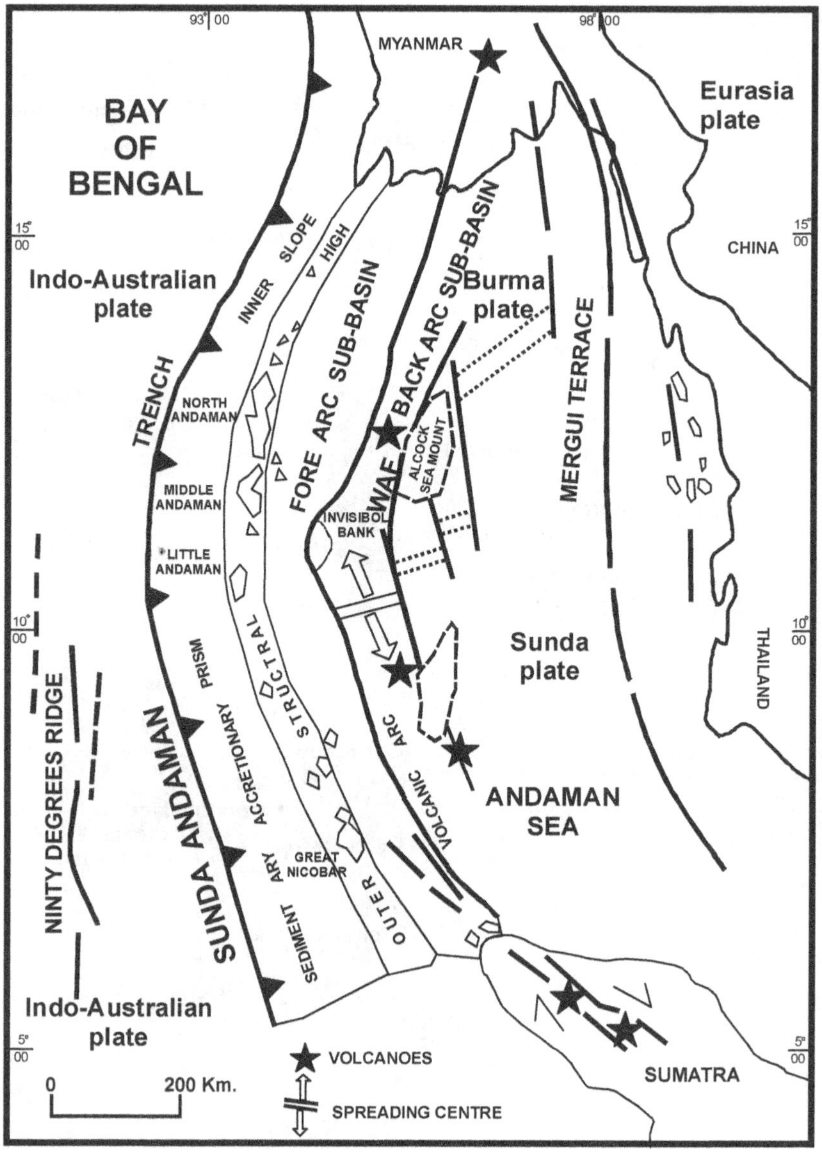

Fig. 12 : Major Tectonic Features

In the Andaman and Nicobar group of Islands, many small tidal estuaries, tidal inlets and the lagoons support dense and diverse mangrove flora. The tidal creeks often are the outlets to the rain-fed stream that flow from the interior and carry silt to the shore to form muddy areas

Chapter 4
WEST COAST OF INDIA

Introduction :

The west coast of India comprises of coast of Gujarat, Maharashtra, Goa, Karnataka and Kerala. All these coastal stretches are affected by specific coastal climate comprising of winds, waves and tidal and littoral currents.

Weather and Coastal Climate :

The climate of the coastal area shows a regular variation on account of the alternating southwest and northeast monsoon. The weather on the coast is therefore more seasonal in nature. December to March is a relatively cool season when the winds are northeasterly (West cost of India: Pilot report by Naval Hydrographic Office). The weather in this season is dry and the cloud cover is very little except in South. April and May are hot months. In this period the winds are light and variable with sea breezes on the coast. June to September is the season of southwest monsoon. Winds on the sea, in this period, are southwesterly and westerly. The winds on the coast, however, are mainly westerly. It is a season of general rains. October and November are marked by light winds. These are considered as transitional months. Occasional tropical cyclones may occur on the Arabian Sea in this period. The period from the end of southwest monsoon to its re-commencement is usually identified as a fair weather season.

In January the region is under the influence of high pressure and the winds are mainly NE or N (the NE monsoon). In July pressure is low over NW India, and SW winds (the SW monsoon) prevail. In the extreme S, there is little seasonal variation of pressure. In the N, however, there is a large difference in the average pressure between the two seasons (Table 10). The predominating wind direction is from NE in January and SW in July.

Considering the specific environment, February to May is treated as Pre-monsoon, June to September as monsoon and October to January as Post-monsoon period on the coast. Rough or very rough seas occur during south-west monsoon. Moderate to heavy swell waves also persist along the coast in this season. Winds on open sea are modified when they approach the coast mainly due to the effect of coastal configuration.

Winds : The SW monsoon affects much of the area from June to August and the wind then blows from the SW quarter with a high degree of constancy. In the NE monsoon (December to March), the wind is not so constant. The predominant direction of the wind is shown for a few representative stations in Table 10.

Between March and May the predominant winds over the sea between latitudes 5° and 20⁰ N change gradually from NE and NW to NW and W. In June, with the onset of the SW monsoon, they become Southwesterly and Westerly and increase in strength; the increase being greatest between latitudes 10° and 20' N. By July, the latter region becomes the windiest part of the area. Over most of the West coast region the wind strength during the NE monsoon is mainly light or moderate.

Local effects cause variations in both speed and direction of the coastal winds. Land and sea breeze effects are usually well developed near the coast except during the SW monsoon.

Table 10 : Weather and Winds

Weather and Waves on West Coast

Parameter Pressure (MSL) mb	kandala	Veraval	Mumbai	Mangaluru	Kochi
April (pre monsoon)	1010	1011	1010	1011	1011
July (monsoon)	1000	1002	1004	1009	1010
November(Post monsoon)	1015	1014	1013	1012	1012
Relative Humidity (%)					
April (pre monsoon)	67	72	73	73	75
July (monsoon)	82	88	85	91	89
November(Post monsoon)	56	51	73	77	78
Wind Speed (knots)					
April (pre monsoon)	15	13	8	6	6
July (monsoon)	19	17	8	5	5
November(Post monsoon)	6	8	5	4	4
Wind Direction & %Days					
April (pre monsoon)	w to sw (34-57)	w to nw (54 - 46)	n to nw (20 - 56)	e & nw (44 & 56)	ne & w (40 & 51)
July (monsoon)	sw (39 - 49)	w (63 to 65)	w (55 to 60)	w to sw (25 - 30)	ne & nw (16 & 38)
November (Post monsoon)	north (34- 51)	s & ne (25 & 54)	ne (40)	e &nw (79 &43)	e& w (37 & 44)

Source : (Naval Hydrographic report on West Coast of India, 1981)

Wind direction and wind speed show definite trends from north to south (Table10). It is found that the waves are westerly to north westerly in pre monsoon period with a speed varying between 3 and 8 knots. The south Konkan experiences winds of 5 to 15 knots in this period. Monsoon is a period of westerly to south westerly waves with a speed varying from 5 to 20 knots along major part of the coast. The wave approaches and the wind speeds can change locally due to considerable refraction as the waves approach the shoreline. Waves become steeper near the shore especially during monsoon.

The width of the surf zone and breaker zone on this coast decreases considerably in fair weather season. The height of breakers decreases and the number of waves in a breaker also decreases significantly. There is hardly any change in wave period from monsoon to fair weather season. In monsoon northern ends of the beaches experience high wave energy. There is a clear shift of energy conditions from monsoon to post monsoon season. In monsoon everywhere the waves are steep breakers with short wave periods. The sediment supply through the rivers is more. These two factors are responsible for increase in the quantum of sediment in waves.

Variation in sea waves and tidal waves, their intensity and frequency, their approach, height and persistence, are the main factors that influence the processes along the West coast. There is a remarkable north-south and seasonal variation in these attributes all along the coast. The variations are site specific within the major regions.

On the basis of wave heights two distinct seasons can be identified, namely, monsoon (June to September) and fair weather (October to May). Wave heights do not normally exceed 2 m in fair weather. In monsoon waves exceeding 5 m height are seen. Long period waves, with a wave period of 10 to 12 seconds, dominate fair weather season. In monsoon, wave period decreases to 3 to 6 seconds. Towards the end of monsoon wave period increases to 10 seconds indicating the arrival of swells. Significant wave height ranges from 3 m to around 9 m. Predominant average wave period is recorded to be in the range of 2 to 16 seconds on this coast (Table 5).

The Nearshore Zone :

The specific nearshore environment on the west coast is due to all the factors mentioned earlier. The breaker zone near the shore is about 200 m wide from June to September and very high breakers are produced in this period. Wave breakers in monsoon are of spilling and plunging type. In fair weather breakers are characterized by low, surging or collapsing waves. The surf zone produced in monsoon is more than 200 m wide and considerably reduces to less than 25 m in fair weather.

The long shore currents measured at a few places show that these currents are significant especially south of 18 degrees north parallel. The currents are south eastwards in monsoon and move with an average velocity of .03 to .6 m / s. In fair weather they are north north- westwards and have a velocity of .05 to .2 m / s. The direction and the velocity of the long shore currents changes locally and they are mainly influenced by the local coastal configuration (Table 6). North to NNW, currents are strongest in October and SE currents are powerful in July. In monsoon the nearshore area is also dominated by rip currents.

Tides :

Table 11 shows the average tidal range at important coastal sites. The coast experiences mixed semi diurnal tides with a tidal range that varies from less than 1 m to more than 12 m. The tidal range gradually increases from south to north i.e. from 0.8 m at Beypore to 12.6 m at Bhavnagar. Tidal currents are very weak along the coast south of Malpe(13.3⁰N /74.7⁰ E) . Here their velocity rarely exceeds 10 cm /s. Strong tidal currents with a velocity of 70 to 90 cm per second are recorded along the coast north of Okha (22.5⁰ N / 69.1⁰ E). The Southern coastal stretch between Beypore and Mangaluru enjoys micro tidal environment where tidal range is less than 2m. The stretch from Mangaluru to Mumbai is macrotidal with a tidal range varying between 2 to 4 m. North of Mumbai the tidal range is usually more than 4m.This overall pattern however shows variations due to the local geomorphology and configuration of the coast at the tidal station. Tidal mean ranges are usually higher in semi enclosed seas and funnel shaped entrances of bays and estuaries and are typically low on the open coast.

Table 11 : Mean Tidal range on West coast

Station	Lat (DD)	Long (DD)	TR (m)
Bhavnagar	21.8	72.2	12
Navalakhi	23.0	70.5	8
Kandala	23.0	70.2	7.2
Hansthal	22.8	70.3	7.1
Navinar pt	22.8	69.7	6.6
Okha	22.5	69.1	4.7
Dahanu	20.0	72.7	4.6
Mira Bhayindar	19.3	72.8	4.2
Mumbai	19.0	72.8	4.1
Mormugao	15.4	73.8	2.4
Karwar	14.8	74.1	2.4
Malpe	13.3	74.7	2.1
Mangaluru	12.8	74.8	1.8
Kozhikode	11.2	75.8	1.8
Kannur	11.8	75.4	1.7
Bhatkal	14.0	74.5	1.6
Kardamam isl	11.2	72.8	1.4
Kollam	8.9	76.6	1.3
Kochi	10.0	76.3	1.2
Minicoy isl	8.3	73.1	1.2
Pallssery	11.8	75.5	1
Beypore	11.2	75.8	0.8

Source : (Tidal gauge stations)

Sea Level Scenario :

A review of studies pertaining to sea level trends along the West coast of India shows that sea level scenario has been the area of debate and controversies since long. There is no dearth of detailed reports and research findings on this issue. It however seems that dependable information on changing sea levels along Indian coast is yet to be generated.

Sea level changes on Indian coast have been dealt by many scholars (Guzder 1975, Dikshit 1975, Kale and Rajaguru 1985, Karlekar 1981, 2001,Vaidyanathan 1987, Merh 1992, Hashimi et al 1995).. Based on such works tentative sea level curves for different regions have been attempted. The average rise in sea level for India is suggested as 0.67 mm / year as against the global average of 1.8 mm / year (Pachuari, 1996).

Practically, nothing is known about the early Pleistocene, and our in-formation for the late Pleistocene and Holocene strandlines is also fragmen-tary (Merh,1992). It appears from the available literature that the West coast of India is better investigated as compared to East coast as regards sea levels. Many scholars have recognised and used various geomorphic features pointing to emer-gence and submergence of coast due to sea level fluctuations.

The Pleistocene :

The Pleistocene sea level picture, on this coast is quite confusing, and it is rather difficult to infer the various levels attained by the sea. It has been suggested that the terminal Pleistocene sea had gone down to almost —150m along the western continental shelf. It is difficult to categorically state anything about the late Pleistocene transgression during the last interglacial. It probably never arose above the present level.

The highest strandline in Saurashtra was around + 25m. Evidences of this Middle Pleistocene high strandline from other parts of the Gujarat coast have also been recorded (Merh, 1992). Along the Kutchchh coast, several geom-orphic features have been taken as indica-tive of this high strandline. The high strandline seen in Saurashtra, is well recognised along the Mainland coast, in the blue-green clay beds in Narmada Formation. Fur-ther south, towards Maharashtra, a well-defined planation surface 20 to 25m high, abutting against the trappean highlands to the east, could also be a fea-ture indicative of this high sea.

The picture of sea level changes since Middle Pleistocene upward is somewhat clear but we do not have any information regarding the Lower Pleisto-cene. Authen-tic data on Lower Pleistocene sea levels is not available along the Indian coast. Information on early Pleistocene marine sediment is also too scanty. The first ma-jor transgression took place in the Middle Pleistocene. Practically all over Gujarat coast evidences of prevalence of continental conditions prior to lower Pleistocene deposition are recorded.

From the available records it appears that there are few evidences of high strandlines during Lower Pleistocene, and it appears that during its major part, the sea level was mainly regressive and fluctuated only marginally.. On the east coast, little authentic in-formation on the older Pleistocene marine strata is available and, deposits of early and middle Quaternary are missing (Bruckner,1989). On the west coast too undoubted marine strata of Lower Pleistocene have not been re-ported by any one. The first major transgression took place only during the Middle Pleistocene and the sea level went up to almost +25m. This high sea level was attained during the 'Great Interglacial Stage'. The Last Glacial

Stage' was a period of regression when the sea level went down to almost —150m. Regressive conditions prevailed up to 11,000 yrs BP. With the advent of Holocene the sea started rising again; between 6000 to 4000 YBP, it at-tained maximum height, which along the Indian coast appears to have been around +6 to +8m. This Holocene rise of sea level was punctuated by a number of still stands or minor regressions (Merh, 1992).

The scientists at National Institute of Oceanography have developed sea level variation history of the last ~14,500 years B.P., for the western Indian continental margin. To generate the sea level variation curve for the west coast of India, they compiled all the dates of past shore line indicator features available between 21°N to 14°N latitude (till south of Saurashtra Peninsula). These dates ranging from 1,500 to 14,500 yr B.P. were then plotted against height/depth from which the dated material was recovered. The sea level curve was then drawn based on geological reasoning and supporting evidences such as presence of terraces, nature of samples dated and inferences of sea level rise from similar stable areas elsewhere (NIO report).

The curve shows that the sea level along the west coast of India was about 100m lower as compared to present, around 14,500 years before and rose to 80m depth around 12,500 years before with a rate of ~10m/1,000 years. It was followed by a quiet period when the sea level remained unchanged for about 2,500 years. From 10,000 to 7,000 years before, sea level rose at a very high rate (~20m/1000 years). After 7,000 years B.P. it fluctuated to more or less the present level (Fig 13).

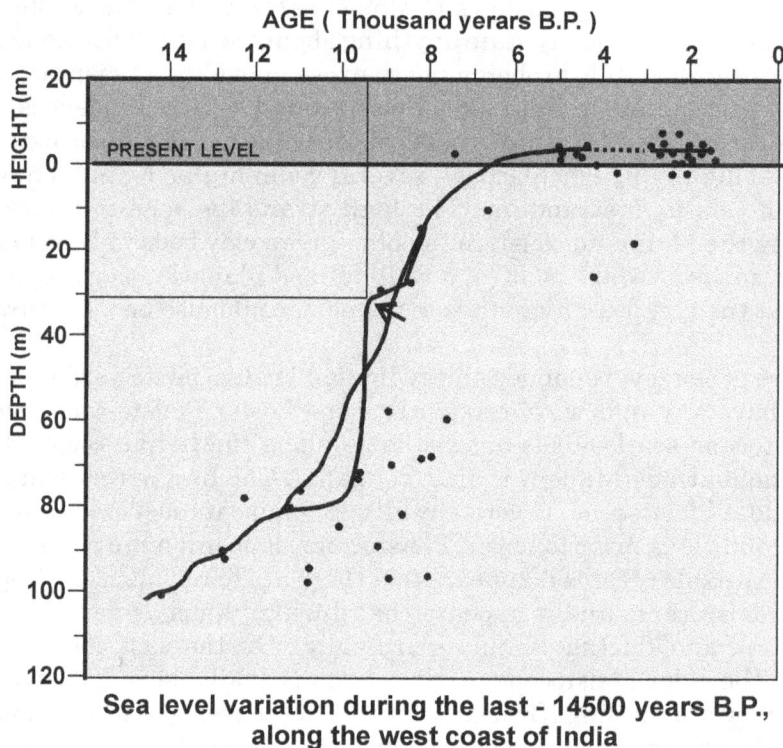

Sea level variation during the last - 14500 years B.P., along the west coast of India

Fig. 13 : Sea level Curve

Studies by NIO group constitute most dependable and authentic data on the continental shelf zone, and the findings provide a lot of information on the behavior of the late Pleistocene-early Holocene sea during its transgressive rise. These workers have recognized four still-stands of sea level indicated by submerged terraces at — 92, — 85, — 75 and — 55m. The radiocarbon ages 9000 to 11000 yrs B.P. indicate formation of these terraces during the transgressive phase of the Holocene sea. Rao(1990) who examined the pisolitic limestones from a depth of 60m came across evidences of these pisolites having formed under semi-arid conditions around 9135 ± 130 yrs BP. Prevalence of inter-tidal to sub-tidal conditions at — 120 to — 150m around 11040 ± 400 yrs BP. have been reported by Nair and Rao(1991).

The Holocene :

The first ever attempt to understand and date sea level fluctuations along the Konkan coast of Maharashtra with the help of radiocarbon age data of beach rocks, was made by Agrawal and Guzder(1972). According to them, between 30,000 to 35,000 yrs B.P. the sea was very much lower; it rose around 30,000 yrs B.P. and this was followed by a regression. The sea rose again around 15,000 yrs B.P. attaining a maximum during the Mid-Holocene, and around 6000 yrs B.P., the sea level was almost the same as present, but in subsequent times, i.e., between 6000 and 2000 yrs B.P., a further rise up to +6m is indi-cated. Kale and Rajaguru(1985) have reconstructed the late Quaternary transgres-sional and regressional history of the west coast with the help of radiometric dates available from other sources along the west coast. The sea level curve prepared by them shows a low (— 138m) sea level around 12000 yrs B.P. They have postulated a rapid rise of level from 12000 yrs B.P. onwards, finally attain-ing the present level around 5000-6000 yrs BP. According to them the sea has always been lower than the present for a large part of the late Pleistocene, while the Holocene sea oscillated above and below the present level several times in the course of the last 6000 yrs.

That the Holocene sea (post-glacial) had arisen several meters above the present level, is amply evidenced by several features along Maharashtra and Gujarat coasts. Occurrence of beach rock (littoral concrete, 'Karat') at heights ranging from 2 to 10m above the MSL along Maharashtra coast all along from Devgad, Ratnagiri in the south to as far north as Dahanu clearly depicts that the Holocene sea had arisen several meters and then regressed to its present level(Karlekar & Raaguru,2012).

This Holocene high sea level is very well indicated in Gujarat (Merh,1992). Along the Mainland Gujarat coast, beach rocks are absent, but instead a good and con-spicuous development of raised mudflats and sandy beaches (with associated stabilized coastal dune ridges) occurring high above the water level (almost + 10m) several kilometers inland, is an obvious evidence of the last high stran-dline, which has now come down to the present level. In Saurashtra, this high sea is represented by semi-consolidated beach rocks and coral reefs. The beach rocks of Saurashtra are identical to those of Maharashtra (Karlekar,2007) and more or less occur at same heights above the MSL. The Holocene sea also appears to have fluctuated in the course of last 6000 yrs, and a marked regression is indicated between 3000 and 5000 yrs BP. (Rajendran,1989).

According to Karlekar and Rajaguru (2012) significant reorientation of the coastline and coastal features such as coastal plains, barrier spits and beaches, estuaries and swamps has occurred throughout the late Holocene (< 4000 YBP) along the entire Maharashtra coast. These configurational changes suggest a sea level change in late Holocene on West coast in general and Konkan coast in particular.

According to Merh (1992), the best example of the Holocene transgression however is the Rann of Kutchchh, which represents remnants of a high sea and provides numerous evidences of strandline regression during historical times. During this period the sea rose to almost +6 to + 8m and indications of this rise are seen as calcareous grits and swash marks on the rocky islands of the Great Ranna, much above the Rann level.

(A) Coast of Gujarat :

Coastal area of Gujarat is largest in the country (28500 sq. km.) and the length of the coastline is about 1600 km. It has the most extensive continental shelf of nearly 164,000 sq km (Figures 1 1, 1 2). Nearly 65,000 sq km of the shelf is within 50m depth contour from the seaboard and the rest is between 50m and 200m contour. These represent nearly 20 and 32 percent of India's coastline and continental shelf.

Gujrat coast is characterized by deltaic drowned irregular prograded coast, straight submerged coast, the spits and cuspate foreland complex, and the mudflat coast. The coast is divided into five regions of which two are bays or gulfs namely the Gulf of Kutch, and, the Gulf of Khambhat, and the remaining include the regions of Rann of Kutch,The Saurashtra coast and the South Gujarat coast (Fig. 14).

Fig. 14 : State of Gujarat

Four major, 25 minor and 5 desert rivers discharge into the coastal waters of Gujarat Coast. The major rivers are Narmada, Tapti, Sabarmati, and Mahi. The major and important minor rivers are listed in Table 12 . Apart from these rivers, the largest river discharging into the North West Arabian Sea is the River Indus. The water from River Indus is considered to be the largest contributor of sediments into the Northern part of the Arabian Sea. The Gujarat coast, from Great Rann to the south Gujarat coast, presents evidence for both emergent and submergent coasts

Table 12 : Rivers Joining Arabian Sea off Gujarat Coast

	River	Catchment area in Gujarat (Sq.Km)	Average Annual Runoff (MCM)	River outflow area
1	Sabarmati	18495	870.52	Gulf of Khambhat
2	Banas	5405	42.86	Little Rann of Kachchh
3	Rupen	2662	143.71	Little Rann of Kachchh
4	Rel	238	9.82	Greater Rann of Kachchh
5	Narmada	11399	6016.31	Gulf of Khambhat
6	Tapi	3837	6627.67	Gulf of Khambhat
7	Mahi	11694	4807.18	Gulf of Khambhat
8	Dhadar	4201	511.9	Gulf of Khambhat
9	Kim	1330	311.16	Gulf of Khambhat
10	Purna	2431	1466.92	Gulf of Khambhat
11	Ambika	2751	1307.86	Gulf of Khambhat
12	Damanganga	495	2938.94	Gulf of Khambhat
13	Shetrunji	5571	34.72	Gulf of Khambhat
14	Bhader	7075	353.33	Saurashtra
15	Machhu	2515	NA	Little Rann of Kachchh
16	Khari	373	NA	Gulf of Kachchh
17	Pur	NA	0.27	Greater Rann of Kachchh
18	Kanakavati	275	0.81	Gulf of Kachchh
19	Gjansar	NA	0.28	Greater Rann of Kachchh

Source : (Planning Atlas of Gujrat 2004)

The Gulf of Kachchh Coast :

On Indian Coast line Kachchh coast is unique as far as coastal processes and geomorphic features are concerned. Total length of Kachchh coastline is about 475 km; out of which 200 km (42.10%) is within Kori-creek, Sir-creek and adjacent area of Deltaic coast. It is 140 km in length, and 70 km wide near the mouth and gets narrowed to about 3 km in at the head of the embayment (Table 13). It enjoys a macro tidal environment with tidal range of 3 m at the mouth to about 8 m at the head; the shape and orientation of the coast, the bathymetry and funnel shaped geometry are the main reasons explaining the amplification of tides. The coastal part to the west of Kori Creek is a part of Sindhu river (Indus) delta. The Gulf is surrounded by arid to semi-arid landmass.

Table 13 : Basic Data - Gulf of Kuchchh

Aspect	Gulf of Kachchh
Area (sq.km.)	7350
Length of Gulf (km)	140
Length of Coast (km)	475
Width near mouth (km)	70
Width near head (km)	03
Average Depth (m)	30
Tidal range (m)	3(near Mouth), 8(near Head)
Mangroves (sq.km)	954
Corals	34 islands with coral reefs
Estuaries	Nil (only seasonal streams)

Circulation in the Gulf is mainly controlled by tidal flows and bathymetry, though wind effect also prevails to some extent. Strong currents normally occur during mid-tide, i.e. 2-3 hrs before and after low and high tides. The spring currents are 60 to 65% stronger than the neap currents. The surface currents are moderate (0.7 - 1.2 m.s^{-1}), but increase considerably (2.0 - 2.5 m.s^{-1}) in the central portion of the Gulf (Gupta 2002).

Tides in the Gulf are mixed, predominantly semi-diurnal type with a large diurnal inequality. The tidal front enters the Gulf from the west and due to narrowing cross-section and resonance, the tidal amplitude increases considerably, upstream of Vadinar. The tidal elevations along the Gulf are illustrated in Table 14.

Table 14 : Tidal Elevations (M) Along the Gulf of Kachchh

Location	MHWS	MHWN	MLWN	MLWS	MSL
Okha	3.47	2.96	1.20	0.41	2.04
Sikka	5.38	4.35	1.74	0.71	3.04
Rozi	5.87	5.40	1.89	1.00	3.60
Kandla	6.66	5.17	1.81	0.78	3.88
Navlakhi	7.21	6.16	2.14	0.78	4.15

Source : (Gupta, 2002) MHWS - Mean High Water Spring; MHWN - Mean High Water Neap; MLWN - Mean Low Water Neap; MLWS - Mean Low Water Spring; MSL - Mean Sea Level.

The distribution of suspended solids in the Gulf is variable and patchy. The deeper regions sustain low suspended load while it is markedly high in the areas close to mudflats and creeks. The inner Gulf areas which are subject to high tidal inundation are highly turbid touching the values as high as 700 mg per litre.

There are many islands and islets in both the Gulf of Kachchh. In the Gulf, there are 42 islands & some islets, covering a total area of about 410.6 sq km. These islands and islets considerably vary in their size. Due to variations in the tidal amplitudes at various locations, the inundation patterns of these islands differs. While many are partially submerged into the water, some of these islands are fully submerged. The inundation pattern of islands is one of the factors which determines the presence of coral reefs and mangroves.

The Northern Coastline of Gulf of Kachchh :

Northern Kachchh coast line is divisible in to five segments: (1).The deltaic coast to the west of Kori Creek, (2).The irregular drowned prograded coast between Lakhpat and Jakhau, (3).The straight coast between Jakhau and Mandvi, (4).The spits and cuspate foreland complex between Mandvi and Mundra, and (5) The wide mud flat coast to the east of Mundra up to the Little Rann (Fig 15).

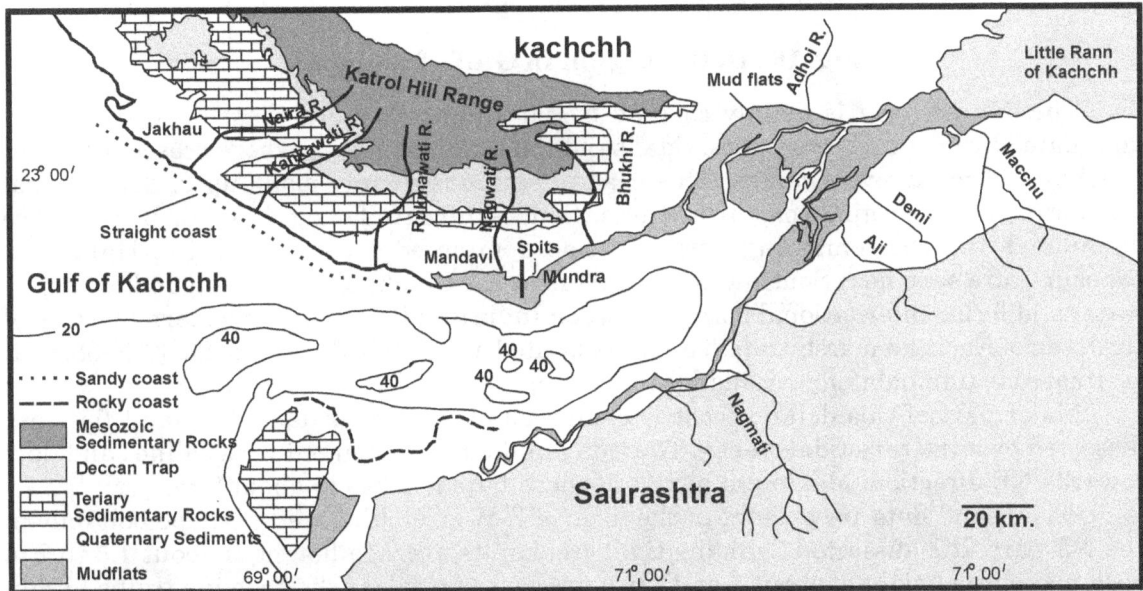

Fig. 15 : Gulf of Kachchh

The large part of the area is covered under shallow sea waters during high tides. So, deltaic coast mainly comprises vast tidal flats in a tide dominated environment. The micro-geomorphic features on the coast consist of tidal mud flats, tidal sand flats, tidal creeks, inshore crescentic barrier beaches and mangroves. The low tidal mudflats in the Gulf occupy about 1590 sq km area and high tidal mudflats occupy about of 586 sq km (Gupta,2002)

(1) The Deltaic Coast to the West of Kori Creek :

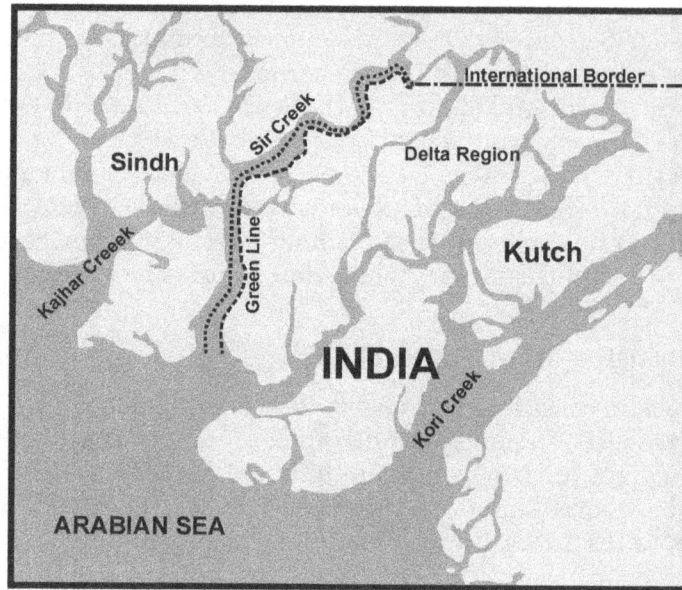

Fig. 16 : Delta Region in Gulf of Kachchh

The deltaic coast is mainly covered by tidal flats. Western coast of Kori creek is dominated by vast tidal mud flats, tidal sand flats, tidal creeks, inshore crescent barrier beaches and mangrove swamps. This coastal sector is mainly made up of depositional landforms and emerging coastal features. Southern coast facing Arabian Sea is wave dominated and the remaining part is a tide dominated coastline(Fig.16). Hard rock exposures are seen near South-western part near Sir Creek. Sir Creek is a 96 km strip of water and forms international border between India and Pakistan on western most tip of the Rann of Kachchh marshlands. The creek located at 23°58 N-68°48 E to 23.96°N-68.87°E is situated in uninhabited marshlands.

Major part of this deltaic coast is composed of tidal mud flats. The mud flats are dissected by numerous tidal creeks. The tidal mud flats are merged with saline sand flats towards NE direction. Maximum height of these flats reaches to about 2 m.

Tidal sand flats are present in the form of E-W stretching wedge shape belt only in the NE part. It is dissected by many tidal creeks. Its average height is about 1.5 to 2 m with maximum height of about 5 m. Dense network of tidal creeks showing trellis to sub-trellis pattern can be seen on the tidal flats.

Seven clusters of crescent shaped barrier beaches are developed to the south, southeast and southwest portion of the deltaic coast which are separated by medium size creeks (Fig. 17). The clusters contain more than 40 barrier islands, where maximum length is approximately 5 km with a width of maximum 250 m. Their height reaches up to 4 m. Dune formation is almost absent. On the backside of some of the barrier beaches, sand flats are produced due to spreading of sand. Cuspate beaches are present at few places (Majethiya et al, 2014). Mangrove patches are seen near south western part covering maximum area of about 8 sq.m.

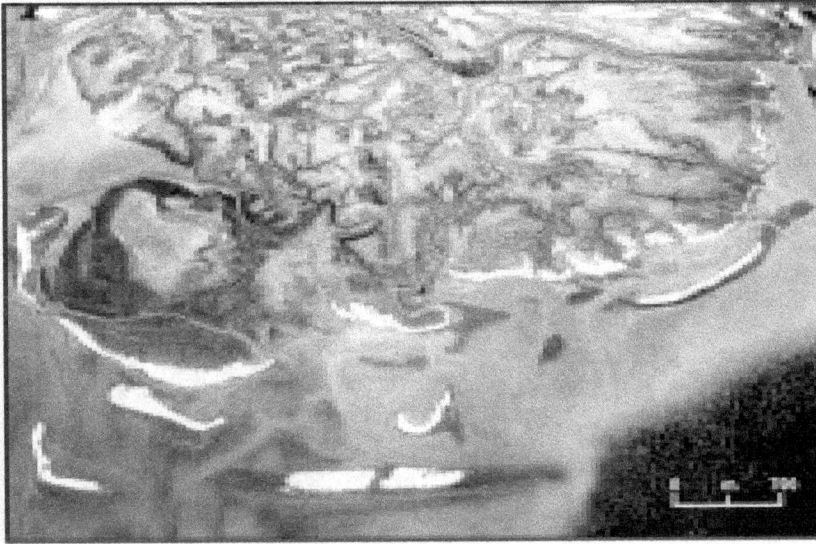

Fig. 17 : Barrier Beaches on Southrn Part Of Delta

(2) The Irregular Coast Between Koteshwar to Jakhau :

This sector has all the features of submergence. A very prominent feature of the Koteshwar (Kori creek) to Jakhau coastal segment is the presence of rocky cliffs almost all along the coastline. The rocky cliffs are 10 – 15 m high and suggest a Palaeo-shoreline. In the earlier period it was a rocky coast which is in contrast to the present day coast characterized by broad intertidal mud flats.This sector is characterized by barrier beaches, fore dunes, tidal creeks, tidal mud flats, parabolic dunes, lagoons, saline sand flats and mangroves (Fig.18)

Fig. 18 : Coastal Features Along Koteshwar to Jakhau Sector

The rocky cliffs are fronted by a narrow zone of raised mudflats. The raised mudflats occur roughly 4 m above the present sea level. They are widest in the area between Akri Moti and Charopadi, where they are overlain by stabilized coastal dunes (Fig.19).

Fig. 19 : Raised Mudflats

The raised mudflats comprise alternating sequences of fine silts and internally laminate organic matter – rich clays. The raised tidal flats are elongated and parallel to the shoreline. It shows a characteristic pattern of distribution of sediments typical of tidal flats. These sediments closely resemble the intertidal mudflat sediments and comprise several horizons. The formation of the raised mudflats and there occurrence at 2-4 m above the present sea level is attributed to tectonic uplift (Bhatt et al, 2016).

There are stabilized coastal dunes in a wide zone all along this coastal sector. The dune morphology is well preserved and comprises several large transverse and parabolic dunes. One such zone is found near Charopadi. At some places the stabilized coastal dunes overlie the raised mudflats while at others they directly overlie the abandoned coastal rocky cliffs. Height of fore dunes ranges from 2 to 4 m. Such dunes are mainly made up of fine well sorted sands. The parabolic dunes are seen on the coast to around 1.15 km inland. The dunes occur from Narayan Sarovar up to Ukir. All the stabilized barrier islands and spits also show this type of dunes. Height of the dunes reaches up to 18 m. Length varies from 300 m to 4 km. On the barrier islands, spits and on the coast the parabolic dunes are covered by vegetation. The dunes are usually oriented in N70°E-S70°W direction.

The rocky cliffs and the stabilized coastal dunes indicate a submerged coast produced by marine flooding during the middle Holocene high sea level. However, the raised mudflats, abandoned rocky cliffs and the incised fluvial terraces among the rivers suggest tectonic uplift of the area in the recent past (Bhatt et al, 2016).

The beaches along the coastal sector are like thin veneer of sandy sediment within intertidal zone. Beaches occur in the form of pockets with a length of few hundred meters and width of 10 to 40 m. At few places they show typical ridge and runnel topography.

There are extensive, gently sloping tidal mud flats in inter-tidal zone. Their width reaches up to 13.5 km where tidal mud flats are formed in form of pockets within sand flats. They are commonly cut across by tidal creeks. Tidal flats are extensively developed on the coast of Narayan Sarovar where their width is 500 m. On the coast near Vayor, Ukri, and Wadsar, tidal flats are mostly protected on the seaward side by active barriers.

There are many saline sand flats having low slope, generally from 2° to 4° on which there is a deposition of comparatively finer sand. Their width ranges from few hundred meters to 13.5 km. Sand flats are also cut across by tidal creeks. Active barrier islands are found towards the coastal side of such sand flats and stable barrier islands are present in central part of the tidal sand flats. The sand flats are composed of mostly calcareous material washed from Kachchh mainland and transported through local river systems (Bhatt et al, 2016).

On the Koteshwar to Jakhau sector, tidal creeks are present wherever tidal flats are present (Fig.18 &19). Most are trellis to sub-trellis with few dendritic and mix trellis-dendritic type of creeks on the coastal tidal flats. While on the tip of Kori creek major ones are trellis with trellis to dendritic sub-branches. The trellis pattern suggests tectonic control over the development of creeks. Creeks near the coast show maximum length of about 16 km and width of about 1.2 km. Such creeks are mostly in their youthful stage with less developed network.

Lagoons are seen at few places on Koteshwar to Jakhau sector (Fig. 18). A cluster of 3 lagoons can be seen near west of Guvar in between parabolic dunes. A lagoon is also formed in and north of Tehra protected by stabilized barrier containing parabolic dunes on the surface. On Koteshwar to Jakhau Sub-segment, mangroves are well developed on vast tidal flats around tidal creeks mostly on eastern side of stabilized barrier islands from Guvar to Jakhau in the inter-tidal zone. Sea cliffs now covered mostly by depositional features, are exposed at few places on Koteshwar to Jakhau coast. The cliffs are visible near Lakhapat, NE of Akri and SW of Akri where their height reaches up to 9 m. Wave cut platforms are absent on the Koteshwar to Jakhau sector(Bhatt et al, 2016). They are probably covered by tidal flat deposits completely.

(3) The Coast Between Jakhau, Mandvi and Kandla Creek :

The coast between the segments Jakhau-Mundra and Mundra-Kandla have irregular and dissected configuration (Fig. 20). The western half, overlooking the open Arabian Sea, trends NW-SE and is dominantly sandy and silty with narrow beaches. The east-west trending coast, that lies inside the Gulf between Mundra and Kandla is made up of extensive tidal flats and merges with Rann of Kachchh to the east. The sites between Mandvi and Kandla creek are occupied with sparse mangroves and rich intertidal mudflats. The total are covered by these mudflats along this stretch is about 115 sq km. Along this coast, the intertidal zone varies from 2 to 5 km between Mundra and Kandla creek.The coast between Jakhau and Mundra has very thin to no mudflats.

Fig. 20 : Part of Jakhau - Kandla Coast

The Southern Coastline of Gulf of Kachchh :

The southern coast of the Gulf from Okha to Navalakhi is occupied by mangroves, corals, intertidal and high tidal mudflats. The inner part of the Gulf shows many salt pans and mudflats. It is a crenulated rocky shoreline with extensive mudflats and the sub tidal zone consisting of tidal channels, shoals, submerged islands, sand bars, coral reefs and mangroves (Fig. 21). Due to vast intertidal area, rocky shores and presence of corals and mangroves, the southern side of Gulf of Kachchh is mostly protected from the problem of coastal erosion.

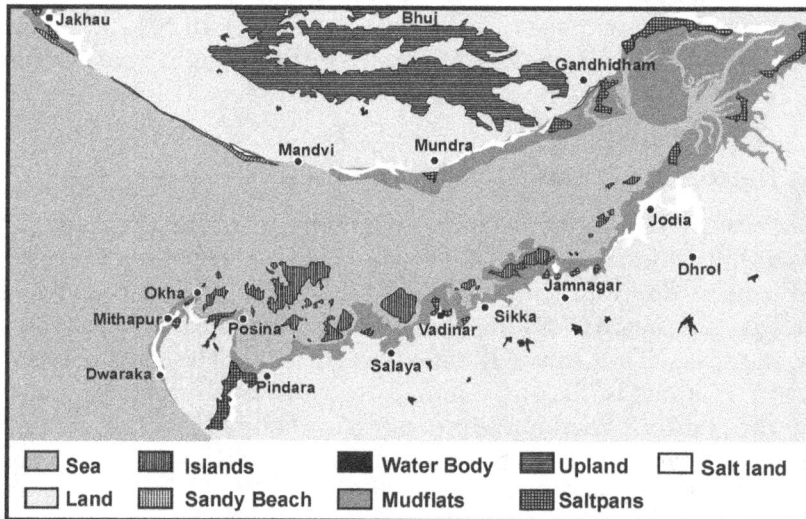

Fig. 21 : Gulf Coast Features

The coral formations of the Gulf of Kachchh are seen between 22° 20' and 22° 40'N and 69° and 70°E along the coast of Jamnagar district. 34 islands out of 42 bordering the southern shore of the Gulf support coral and coral reefs. The age of reefs near Salaya and Okha is reported to be 5240 YBP and 45,000YBP respectively (Gupta, 2002). Pirotan island is the northern limit of the coral growth. Along the southern coast, the intertidal zone extends to about 1 km in the outer Gulf, 2 to 3.5 km in the central Gulf and 3.5 to 5 km in the inner Gulf.

The Gulf of Khambhat Coast :

The Gulf of Khambhat coast is indented by estuaries and consists of mudflats, dunes, and beaches. Here mudflats are seen at different levels and paleo-mudflats are related to sea level regression.

It is a large estuarine embayment that opens southwards into the Arabian Sea. It is more than 200 km long and up to 70 km wide at its southern end, 190 km wide at its mouth between Diu and Daman narrowing to 24 km towards its head (Table 15) (Fig. 22). The gulf receives many rivers, including the Sabarmati, Mahi, Narmada (Narbada), and Tapti. Its shape and its orientation in relation to the southwest monsoon winds account for its high tidal range (12 meters) and the high velocity of the entering tides. The Gulf is known for its extreme tides, which vary greatly in height and run into it with amazing speed. At low tide the bottom is left nearly dry for some distance below the town of Khambhat. There are two islands bets in the Gulf, the Piram bet near Bhavnagar and Alia bet near Narmada estuary close to Bharuch and Ankleshwar towns. Estuaries are mainly formed along the eastern bank of the Gulf.

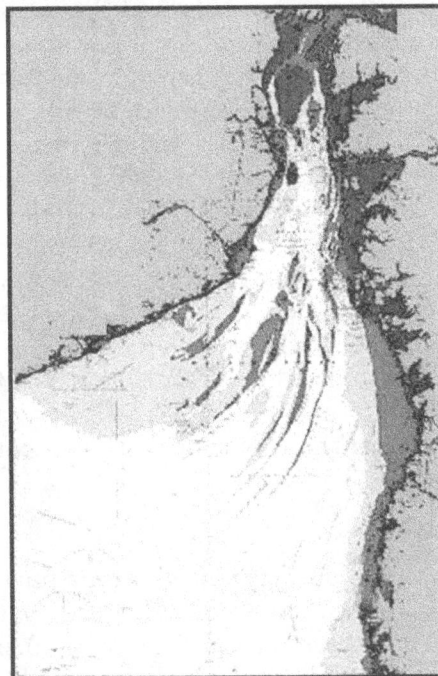

Fig. 22 : Gulf of Khambhat

Table 15 : Basic Data on Gulf of Khambhat

Aspect	Gulf of Khambhat
Area (sq.km.)	3120
Length of Gulf (km)	200
Length of Coast (km)	370
Width near mouth (km)	70
Width near head (km)	24
Average Depth (m)	40
Tidal range (m)	3(near Mouth), 12 (near Head)

Tidal sand bars and tidal sand ridges are extensively developed in the macro tidal Gulf of Khambhat.

The inner and outer regions of the gulf are characterised by the development distinct tidal sand bodies with discrete geometries and dimensions (Saha *et al*,2016). These tidal sand bars occur in the estuary mouths and within the tidally inûuenced ûuvial reaches of the rivers ûowing into the gulf. The height of these sand bars is in the range

1to3 m. Due to high tidal ranges and bidirectional ûow the sand bars do not achieve signiûcant heights. The Gulf of Khambhat acquired the present conûguration in the last few thousand years since the Pleistocene sea-level low strand (Fig.23). Tides in the gulf at the estuary mouths are semi-diurnal with a spring tide range of up to 10 m making the gulf a macro-tidal estuarine environment(Saha *et al*,2016).

The tidal range at the Narmada Estuary mouth varies from 5 to 10 m . The semi-diurnal tides in the Gulf of Khambhat amplify threefold from the mouth to the head of the with maximum recorded current speeds varying from 1.4 m/sec in the open coast to 3.2 m/sec in the upper reaches of the Gulf of Khambhat). During the southwest monsoon the west coast of India the region is characterised by larger waves with relatively calm sea conditions during the rest of the year. Wave height during the monsoon varies from 0.1 to 2.9 m in the inner gulf to up to 6 m in the outer area.

Some of the vast stretches of mud flats occur along the Gulf of Khambhat. There is quite a vast stretch of mud flats near Ghogha in Bhavnagar.on western bank of the Gulf. Marshy coasts are found near the estuarine systems located mostly along the eastern side of the Gulf of Khambhat.

The Little Rann of Kachchh :

Rann of Kachchh is a saline desert for the larger part of the year and is further divided into the Great Rann and the Little Rann.They are special types of wetlands. Little Rann of Kutch occupies south eastern part of Rann of kachchh. It is a saline-inundated regime, partially seasonal and partially for the whole year (Fig 24).

The area gets flooded by both the tidal water ingression and freshwater poured by

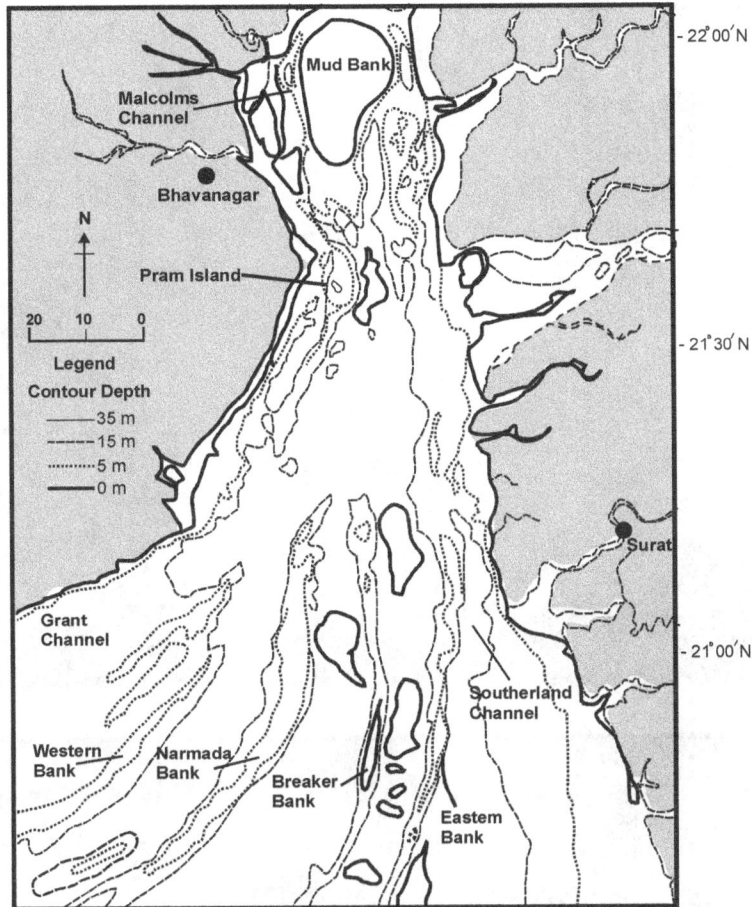

Fig. 23 : Bathymetry of Gulf of Khambat

the seasonal rivers during the monsoon season. For the remaining part of the year, the region is a vast expanse of sun baked mud and sand. The area therefore experiences a seasonal reversal of geomorphic processes. Geologically, this area was a part of oceanic floor and has immerged in the recent past. The saline water intrusion is a result of its distinctive terrain configuration. The surface of Rann is at or slightly above sea level and possesses a monotonously plain character with some outcrops in the form of sandstones capped by basalts, resembling island in the mid of the Rann(Gupta,2014). During southwest monsoon, storm tides force water from the Arabian Sea over the flat surface of the Rann. Rainfall is fairly low, so that as the water recedes and evaporates, it leaves behind a crust of halite and gypsum crystals which grow in the clay and sands. Its terrain is thus a monotonously saline flat surface produced by annual inundation

The Saurashtra Coast :

It shows numerous cliffs, islands, tidal flats, estuaries, embayments, sandy beaches, dunes, spits, bars, bays, marshes, and raised beaches at some places.

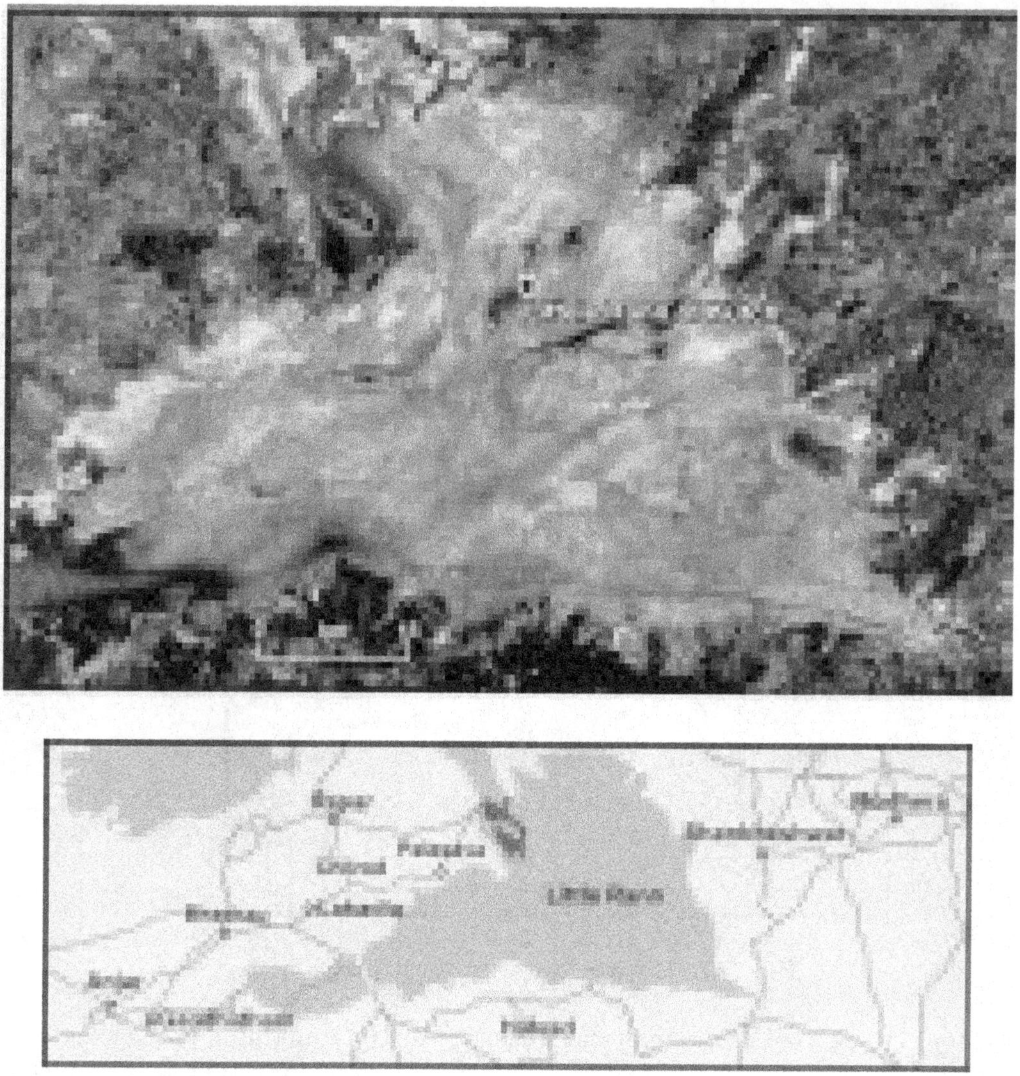

Fig. 24 : Little Rann of Kachchh

Geomorphic expression of land sea interaction is preserved in the form of abandoned cliûs, marine terraces, shore platforms and marine notches along the southern Saurashtra coast. These featuresgive an idea of the magnitude of sea level changes during late Quaternary. Coastal cliffs and shore platforms are important geomorphic features of coastal areas of Saurashtra. These features are composed of medium to coarse grained carbonate sand and are designated as "Miliolitic limestones" that range in age from Middle to Late Pleistocene(Marathe *et al* 1977,Marathe 1981, Baskaran et al 1986).

The coastal segment lying between Kodinar and Babarkot is dominated by the bioclastic carbonate deposits (miliolite) of late Quaternary age. There are two distinct types referred to as miliolite limestone and shell lime-stone which form coastal sheets, beach ridges and associated aeolianites along with patchy occur-rences of dead coral reefs (Bhatt *et al* 2006). Stratigraphically, they range in age from early Middle Pleistocene to Late Pleistocene

The coast falls under micro-tidal to meso-tidal regime with a tide range from 1.8 m at Veraval to 2.54m at Pipavav Port (near Jafrabad). A 10 m high cliff and associated head land at Babarkot (20°52' N; 71°24' E) shows morphological features like raised marine benches and distorted tidal notch. A promi-nent raised shore platform 3.5m above sea level and an associated notch having 1.5 m roof height have been observed. The miliolite limestone cliff at Jafrabad light-house exhibits a distinct shore platform at the height of 4 m.

A cuspate beach abutting against a moderate cliff could be observed near Sarkeshwar temple (20°50' N; 71°17' E) (Bhatt *et al* 2006). The most prominent sea cliff about 14 m in height has been recorded near Nagowa (20°42' N; 70°54' E) on Diu Island (Fig. 25). Long sandy beaches are found along the Junagadh and Porbandar coast.

Fig. 25 : Landforms on Diu Coast

The South Gujarat coast :

This part of Gujrat coast is relatively uniform and is indented by a series of creeks, estuaries, marshes, and mudflats. It stretches from 20°08' to 21°30' N latitudes and has a length of about 125 km. (Fig.26) The maximum tidal height and significant wave height increase from northern side i.e. near Hajira to southern side i.e. near Umergaon. The maximum tidal height near Hazira and Valsad are 5.78 m and 7.77m respectively where as the maximum significant wave height near Hazira is 0.97 and near Valsad is 2.47m.

In Gujarat sandy beaches constitute about 28% of total coastal length and mainly occurred along South Gujarat coast near Umbargam in Valsad district. Narrow sandy beach is present between Mindhola and Purna rivers and extens up to Daman. Mudflats, marsh and mangrove vegetation are found along the estuaries of the Mindhola, the Purna, the Ambica, the Auranga and the Damanganga (Mahapatra et al,2013). Numerous small tidal creeks are also found along the study area. South of Auranga estuary, the coast is rocky.

It has been seen that about 69.31 % of the South Gujarat coast is eroding, about 18.40 % is stable and remaining 12.28 % of the coast is accreting (Mahapatra *et al*,2014).

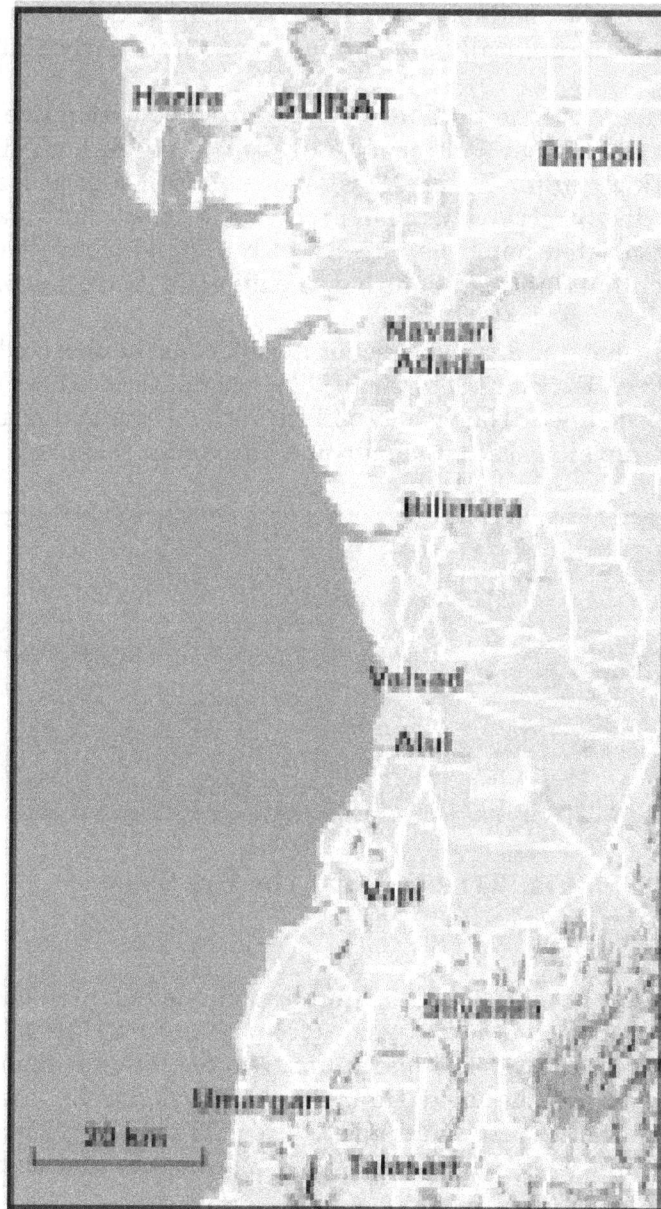

Fig. 26 : South Gujarat Coast

Their study showed that high erosion area are mostly restricted to the coast along Umergaon (near Fansa, Maroli, Nargol, Varili river mouth, Umergaon light house) and Pardi (Kolak, Udwara)Taluka in Valsad district. Stable coastal length of the study area is 21.59 km and was found in Nani Dandi and near Onjal. High accretion (3.70 %) was seen near Hajira. The main causes of coastal erosion along the coast are strong tidal currents accompanied by wave action.

(B) Coast of Maharashtra :

A 500 km long coastline (720 km long including estuarine and creek coast) and a narrow coastal plain stretching from north to south along the western boundary, is a distinct physiographic region of Maharashtra (Fig. 27). The region is traditionally known as Konkan and is a land of plateaus, plains and hills. It is separated from upland Maharashtra by a west-facing escarpment of Sahyadri Mountains or Western Ghats.

Konkan extends from Damanganga River in the north to Terekhol River in the south. The width of this coastal belt is not uniform and varies from 40 to 50 km all along the region. Administratively it comprises of Palghar, Thane, Raigad, Ratnagiri and Sindhudurg districts and Mumbai. (Fig. 27)

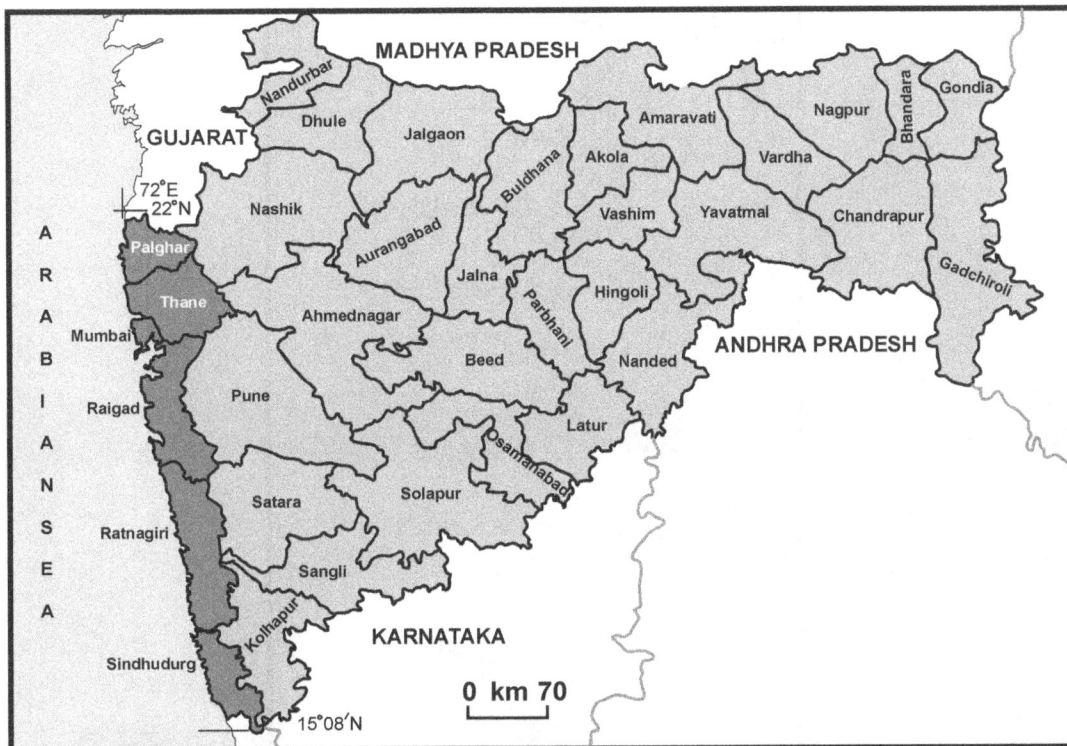

Fig 27 : State of Maharashtra

Physiographic Divisions of Konkan Region :

On the basis of certain characters like lithology, geomorphic configuration, nature of hinterland and climate, Konkan region of Maharashtra can be divided in to North Konkan, Middle Konkan and South Konkan. Dikshit (1986) also suggests this type of a division based on Physico-cultural characteristics.

North Konkan: The coastal belt from Bordi, Dahanu to Karanja (Fig.28) is included in this sector. The region is mainly characterized by forested hills and a small plateau of a height of about 350-m ASL.The drainage network comprises of rivers like Surya, Tansa, Vaitarna and Prinjal, Ulhas, Kalu and Bhatsai. Courses of rivers like Vaitarana clearly indicate a control of lineaments.

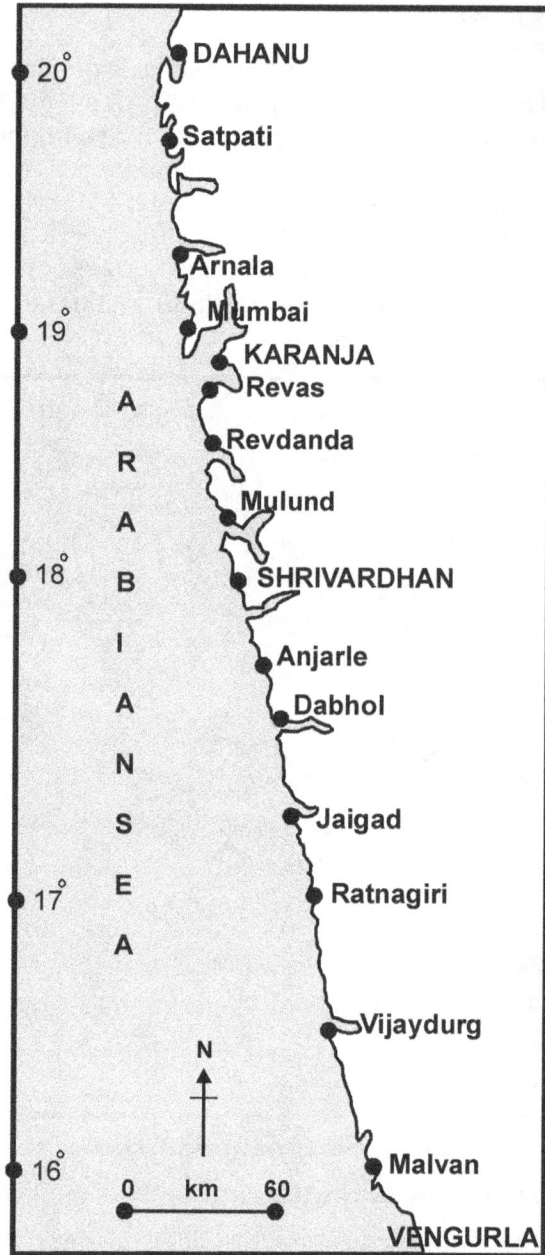

Fig. 28 : Konkan Coast

Narrow plains invariably border the river channels. Ulhas plain, covered by brown silt, is according to Dikshit (1986) a marine planation surface in early quaternary. North-south oriented, forested coastal hill range is an important feature of North Konkan. The entire region is an area covered by the Deccan Trap rocks with intertrappean beds. North Konkan receives an annual rainfall of 1500 mm and more.

Middle Konkan: The coastal belt from Karanja to Shrivardhan can be identified as Middle Konkan. This region also comprises of forested hills but it is relatively more flat. Amba, Kundalika, kal, and Savitri are the main rivers and show a distinct control of lineaments. The central part of the east - west belt of this area is hilly with height ranging from 300 m to 500 m. Dikshit (1986) identifies most of these hills as residual hills, suggesting a planation surface at about 550 m ASL. Basalt is the major lithological formation in the area.

South Konkan: This is the longest stretch of Konkan between Shrivardhan and Vengurla, covered by thick laterites .The laterites have molded the personality of the region when compared to North and Middle Konkan. Barren lateritic plateaus, deeply entrenched river channels and the piedmont plains at the foot of the Sahyadrian escarpment are the significant land facets, which reflect the impact of lithology (Karlekar, 1981). The plateau is 150 to 200 m high and the cover of laterite is 8 to 12 m thick. Vashshthi, Shastri, Kajavi, Muchkundi, Arjuna, Vaghotan, Gad and Karli are the major rivers in the area. South of Tarele, the cover of laterites is progressively reduced. The granitic and gneissic formations influence the landform development, especially around Kankavali, Kudal, Sawantwadi, Malvan and Vengurla. The area is rich in minerals. Iron, manganese, bauxite and the silica sand are the important mineral reserves of the area.

Divisions of Coastal Sector :

The coastal sector, characterized by plains, shoreline terraces, sand dunes, cliffs, numerous sandy pocket beaches, tidal inlets, creeks and estuaries shows a great amount of variability from north to south. The landward margin of the coastal hinterland can be identified as the north – south hill ranges in the central part of Konkan, roughly parallel to shore. The hill range, although not continuous, forms a watershed between coastal streams and the scarp foot plains to its east. The coastal strip shows a varying width. It is wider in the north than in south.

On the basis of the impact of tidal incursion and the tidal range in spring and neap, Konkan can be divided into macro, meso and micro tidal regions. Table 16 gives the classification of Konkan based on these aspects.

Table 16 : Coastal Divisions of Konkan

Stretch	Macrotidal Coast Dahanu to Revas	Mesotidal Coast Revas to Ratnagiri	Microtidal Coast Ratnagiri to Redi
Spring tide range	>3.5 m	3.5 to 2 m	< 2 m
Neap tide range	> 2 m	2 to 1.5 m	< 1.5 m
tidal Incursion	40 km	25 km	20 km
(Source : Author)			

Coastal Climate :

Wind direction and wind speed on this coast show definite trends from north to south. It is found that the waves are westerly in pre monsoon period on middle and south Konkan coast and north westerly on northern coast with a speed varying between 3 and 8 knots. The south Konkan experiences winds of 5 to 11 knots in this period. Monsoon is a period of westerly to south westerly waves with a speed exceeding 10 knots along major part of Konkan coast. The wave approaches and the wind speeds can change locally due to considerable refraction as the waves approach the indented shoreline. Waves become steeper near the shore especially during monsoon

On the basis of wave heights two distinct seasons can be identified, namely, monsoon (June to September) and fair weather (October to May). Wave heights do not normally exceed 2 m in fair weather. In monsoon waves exceeding 5 m height are seen, especially along south Konkan coast. Long period waves, with a wave period of 10 to 12 seconds, dominate fair weather season. In monsoon wave period decreases to 3 to 6 seconds. Towards the end of monsoon wave period increases to 10 seconds indicating the arrival of swells.

The breaker zone near the shore is about 200 m wide from June to September and very high breakers are produced in this period. Wave breakers in monsoon are of spilling and plunging type. In fair weather breakers are characterized by low, surging or collapsing waves. The surf zone produced in monsoon is more than 200 m wide and considerably reduces to less than 25 m in fair weather (Karlekar2014, Karlekar & Rajaguru 2012).

The long shore currents measured at a few places show that these currents are significant especially south of 18 degrees north parallel. The currents are southeastwards in monsoon and move with an average velocity of 30 to 40 cm / s. In fair weather they are north northwestwards and have a velocity of 8 to 20 cm / s. The direction and the velocity of the long shore currents changes locally and they are mainly influenced by the local coastal configuration. North to NNW, currents are strongest in October and SE currents are powerful in July. In monsoon the nearshore area is also dominated by rip currents.

It has been reported that Maharashtra coast experiences negligible annual long shore sediment net transport and the direction of annual net transport is towards south (Kunte et al 2001). At Mirya bay, southward drift is estimated to be between 75,000 m³ and 150,000 m³ per year. The annual net transport direction varies along the west coast. It is southerly at Ratnagiri and northerly along Vengurla and Goa (Kunte et al 2001). Based on the field measurements estimated long shore sediment transport rates at Ratnagiri, Ambolgarh and Vengurla are reported to be 119,000 m³, 190,000 m³ and 53000 m³ per year respectively (Kunte *et al.* 2001).

The Konkan coast experiences semi diurnal tides with a tidal range that varies from less than 2 m to more than 3.5 m. The tidal range gradually increases from south to north i.e. from 1.5 m at Vengurla to 5.4 m at Valsad (Table 17). Tidal currents are very weak along the south Konkan coast. Here their velocity rarely exceeds 10 cm /s. Strong tidal currents with a velocity of 70 to 90 cm per second are recorded along the north Konkan coast.

Table 17 : Tidal Range at Few Important Places

SITE	LOCATION ° ′ ° ′	SPRING TIDE (m)	NEAP (m)
Valsad	20 38, N / 72 53, E	5.4	3.6
Dahanu	19 58, N / 72 43, E	4.1	3.4
Satpati	19 43, N / 72 42, E	3.6	2.6
Arnala	19 27, N / 72 45, E	3.9	2.0
Vasai	19 19, N / 72 48, E	3.8	1.9
Mumbai	18 55, N / 72 50, E	3.6	1.9
Karanja	18 55, N / 72 56, E	3.6	2.1
Revas	18 49, N / 72 57, E	3.6	2.0
Revdanda	18 33 ,N / 72 56, E	2.5	1.8
Murud	18 19, N / 72 58, E	2.5	1.8
Bankot	17 58, N / 73 03, E	2.0	1.5
Dabhol	17 35, N / 73 11, E	2.7	1.6
Jaygad	17 18, N / 73 14, E	2.7	1.5
Ratnagiri	16 59, N / 73 18, E	2.1	1.3
Musakagi	16 37 ,N / 73 20, E	1.8	1.4
Vijaydurg	16 33, N / 73 20, E	1.8	1.3
Devgad	16 23, N / 73 23, E	1.9	1.3
Malvan	16 03, N / 73 28, E	1.5	1.2
Vengurla	15 51, N / 73 37, E	1.6	1.2

Source : *(Naval Hydrographic Report on West coast of India, 1981)*

Changing Sea levels :

Konkan is an excellent region which has preserved marks of former sea-levels. The rapid development of road links to remote and inaccessible areas has helped in the search for the evidences of former sea-levels on the coast. Degree of preservation of morphological features is relatively better than that of the sedimentary features. Shore platforms, notches, sea caves and cliffs some distance inland and the loose beach-dune deposits, consolidated to produce littoral concrete/beach rock/aeolinites in the nearshore to backshore zones, provide convincing evidences for plaeo sea-levels on this coast. There appears a prefect association between the configuration of the coast and the morphological and sedimentary features, suggestive of ancient and historic sea-levels. The indented and rocky nature of the coast has helped in the preservation of the evidences, especially the morphological features.

A variety of morphological features and fossil sediments on Konkan coast of Maharashtra display a transgressive/regressive record. Transgression of sea in early Holocene was followed by a regression which was possibly interrupted by minor advance of sea-level.

The modern sediments, beaches and dunes of the coast, are backed by old, fossilised beaches and dune terraces and cliffs. The fossil deposits are more or less parallel to the shore. The deposits are calcareous, sandy and shelly in nature and occupy varying topographic positions in the area. Many coastal villages are situated on such fossil dune and beach ridges.

The fossil deposits on Konkan coast are variously termed as beach rocks, aeolinites and karel (Guzder, 1975; Dikshit, 1975; Karlekar, 1981). Their occurrence is patchy and frequently they are concealed under the modern coastal alluvium. The fossil beach and dune ridges can be located between 20 to 700m from high tide limit. Their rarely exceeds 3.5m ASL.

The occurrence of deposits in variable settings such as creeks and streams creates complications and attributing them to specific sea-levels becomes difficult. Moreover, they are heavily eroded and thin. Therefore it is not easy to estimate their extent at the time of their formation. The morphological features such as ancient cliffs, caves, shoreline terraces also give clues to the vertical displacement of shoreline in the area. The study of near shore features as the indicators of palaeo sea-levels is more meaningful in regions like Konkan which are mainly regressive in nature.

It is necessary to identify and correlate the evidences indicative of minor fluctuations affecting horizontal distance of not more that 8 to 10 km from the present shore. Once a feature was suspected of being produced by the erosional or depositional work of the sea, its lithological composition, shape, orientation, relative height, sediment composition and its distance from the present shoreline was recorded. The features studied include fossil beach and dune ridges, tidal basins, shoreline terraces, wave cut notches, cliffs, caves and shore platform.

The shoreline terraces at Palghar, Mahim, Ravas, Alibag, Nandgaon, Shrivardhan, Dabhol, can be easily indentified in the field. The beach rocks from a variety of geomorphic settings such as creeks and dunes can be observed at Kashid, Nandgaon, Dive Agar, Adgaon, Sarve, Mirya, Ganeshgule and Malvan.

According to Karlekar and Rajaguru (2012) there are many conflicting field evidences of Holocene sea level changes along entire Konkan coast of Maharashtra. The reported field evidences for Holocene sea level changes are still unequivocal and that detailed topographic and stratigraphic studies of the coastal area would be required to elucidate the Holocene sea levels. These sea level changes were mainly recorded as configurational changes. The identification and mapping of coastal features along the coast reveal that during the late Holocene this coast was characterized by barrier spits, open inlets and estuaries. Wide dune systems comprising of fore and back dunes, mangrove swamps, abandoned and realigned spit bars and littoral terraces found almost everywhere on this coast suggest some degree of reorientation and configurational change in the late Holocene.

The estuarine systems on this coast probably matured during the late Holocene, with progressive sedimentation and seasonal inlet closure, leading to the dominance of mangrove swamps in the past 1000 years. Due to change in sea level and availability of sediments to coastal areas the littoral sediment traps got enriched with clastics which helped in realigning the coast in later part of Holocene. Infilling of tidal inlets, especially the estuaries, formation of spits and littoral terraces are all related to these factors.

Evidences of episodic high wave energy events are reflected in the form of abandoned and realigned spits on this coast. Beach dune sediments along the former coast were lithified into beach dune ridges sometime after 3800 YBP. The prior estuaries maintained configuration at least until 3000 YBP (Karlekar & Rajaguru,2012). After this progressive slow infilling of estuaries occurred. Recently after 700 YBP, a change from estuarine sedimentation to swampy sedimentation occurred, which enabled rapid mangrove colonization within estuaries. This can be linked to a slow and slight lowering of sea level

The most notable fossil rock formation found on Konkan coast is creek beach rock (CBR) at maximum distance of 2200 m from the mouth of creeks. Here, these deposits are like cemented, calcareous sandstones. They are very brittle and can be crushed in fingers. The major part of the material consists of shells and fine sand. In tidal inlets CBR can be found within an elevation range of 0.5 to 2.5 m (Table 18). Not a single rediocarbon date is available for CBR. One can, however, infer an age of 500 to 1600 YBP from the sea-level curve prepared for later part of Holonce, from the dates available for other fossil deposits.

The intertidal beach rocks (ITBR) are exposed at low tides. They show relatively better cementation of shells and fine sands. The cemented material is angular to subangular. At low tide ITBR can be seen 8 to 10 m inland on the upper beaches. These rocks suggest a recent fall in sea-level.

Fossil deposits of beach and dune origin are geomorphically very significant on Konkan coast. The thickness of these deposits ranges from 2 to 8m. They consist of a homogeneous mass of shells and fine sand. As a rule proportion of marine shells increases towards the top of the deposit. Both ocur in a wide elevation and distance range, 0.5 to3.6 m and 20 to 700m, respectively. They suggest a higher sea-level in this part of west coast of India 1400 to 3800 YBP.

Roots of mangroves, buried in a 20 cm thick layer of mud, were found at Revas. The radiocarbon date from the area suggests a historic period of 520 to 680 YBP. The old sea cliffs, shoreline terraces, shore platforms and defunct tidal basins are the morphological features that occur at varying distances and heights in the area. The features are conspicuous and certainly related to former sea-level in the region. Due to the absence of any dateable material, these features cannot be related to specific strandlines. However, considering the elevation range, shoreline terraces (5 to 6 m ASL) and defunct tidal

Table 18 : Fossil Beach Dune Deposits on Konkan Coast

Place	Location	Feature	MXrrial	Dis. (m)	Elevation(m)	Age(YBP)	Reference
I . Navapur	19°46'N	CBR	shells and fine Sand	2200	1.5	-	-
		FBR	Sand	700	2.O	-	-
		FBR	Shell. Sand	150	1.2	2080+50	Karlekar (1981)
		ITBR	Littoral Concrete	8	-	-	-
2. Palghar	19°44'N	CBR	Shells	150	0.5	-	.
3.Uran	18°50'N	FDR	Sheets, Shingles	1000	3.5	1800+50	Bruckner (1989)
		FDR	Sand	80	0.5	1470+55	Bruckner (1989)
		ITBR	Littoral Concrete	10	-	-	-
4. Karaja	18°49'N	ITBR	Littoral Concrete	10	-	-	-
5. Revas	18°49'N	Buried	Rotten Roots	12	0.7	600+80	Karlekar (19%)
6.Dharamtar	18°47'N	Marine Fossil	Oyster Shells	-	-3.0 RiverBed	2410+95	Bruckner (1989)
7. Avas	18°45'N	FDR	Shells, Sand	220	22	-	-
8. Korlai	18°30'N	FBR	Littoral Concrete	9	-	2410+95	Agrawal & Guzdcr (1972)
9.Barshiv	18°20'N	FBR	Sand,Shells	50	1.0	-	-
10.Nandgaon	18°19'N	FDR	Sand	450	3.0	-	-
11.Velas	18°10'N	CBR	Fine Sand. Shells	900	1.5	.	-
		FBR	Shells	20	0.8	-	-
12. Divaegar	18°08'N	FBR	Shells	250	3.0	2350+60	Karlekar (1996)
13.Shekhadi	18°06'N	FBR	Shells	50	3.6	3800 + 90	Deswandikar (1993)
14. Valvati	18°50'N	FBR(I)	Shells	200	2.5	2260+60	Bruckner (1987)
		FBR(II)	Shells	200	1.5	2410+60	Bruckner (1987)
15. Kelshi	17°55'N	FBR	Shells	500	5.0	-	-
16.Mirya	17°01'N	FBR	Shells	400	5.9	2800+10	Agrawal & Guzder (1972)
		FBR	Sand,Shells	450	6.0	2305+95	Agrawal & Guzder (1972)
17. Ratnagiri	17°N	FBR	Sand,Shells	450	3.5	2305+50	Bruckner (1989)
18.Agargule	16°47'N	CBR	Sand,Shells	40	2.5	-	-
19.Vijaydurg	16°33'N	FDR	Sand	60	1.5	2350+60	Karlekar (1981)
20.Kolamb	16°09'N	CBR	Sand	150	2.0	-	-

CBR	-	Creek Beach Rock
FBR	-	Fossil Beach Rock
FDR	-	Fossil Dune Ridge
ITBR	-	Intertidal Beach Rock
FD	-	Fossil Dune

Source : *Author*

basins (1 to 1.8 m ASL) can be linked with 2000 and 1500 year old higher sea-level in Konkan

The features identified along the coast occur at various distances inland from the present shoreline. Shore platforms are seen up to a distance of 35 m and up to an elevation of 1.5 m from the high water line (Table 19).

Table 19 : Morphological Evidences of Shoreline Movements on Konkan Coast

Feature	Distance Range from HWL (m)	Elevation Range ASL(m)
Shoreline Terraces	6 to 2400	4 to 6
Shore Platforms	upto 35	0.1 to 1.5
Abandoned sea cliffs	30 to 3000	4 to 9
Defunct Tidal Basin	1500 to 1800	1 to 1.8

Shoreline terraces covered with coastal alluvium can be traced up to 2.4 km and occur between an elevation range of 4 to 6m. The limit of defunct tidal basins, especially at Navapur, Velas Deveagar and Valvati, can be indentified 1.8 km inland. Such basins which are defunct, however, do occur frequently on the coast. Abandoned sea cliffs, can be seen from 30 m to 3 km inland and their height rarely exceeds 9m. The cliffs at Nandgaon, Harihareshwar, Ratangiri and Vijaydurg are the examples of such abandoned cliffs. The fossil dunes and fossil beaches are frequently seen within a distance range of 20 m to 90 m depending on the nature of coastline. In a region characterised by creek or estuary and a wide flat plain, they are usually detected to a greater distance inland than in a rocky-headland region. Fossil mangroves are very rare and occur within 10m from the present sea. Most of the fossil deposits of dune and beach origin are dated to 1500 and 3800 YBP. They were found to occur from 1 m to 6 m elevation from present sea-level. There appears a perfect association between the configuration of the coast and the various features on the coast. the terraces, fossil beach ridges, the fossil dune ridges are essentially found bordering tidal basins. The extent of these features is seen to be governed by backshore relief. The rocky shores and headlands are the sites where ancient shoremarks and notches are found. There are ample evidences of morphological changes which may be due to neotectonism and sea-level fluctuations in the area. Such evidences, in all probability, are indicative of shoreline changes in mid to late Holocene.

Landforms on Konkan Coast :

All along the region the shoreline is broken by frequent headlands and promontories, which are the sites of steep sea cliffs, beautiful sandy pocket beaches, drowned river valleys, small tidal inlets, river creeks and creek lets. (Fig. 29) One is thrilled by an almost regular sequence of headlands and tidal inlets. Narrow, flat and low shoreline terraces covered with a thin apron of coastal alluvium border the tidal inlets. These land facets have contributed immensely to distinctiveness of the Konkan coast.

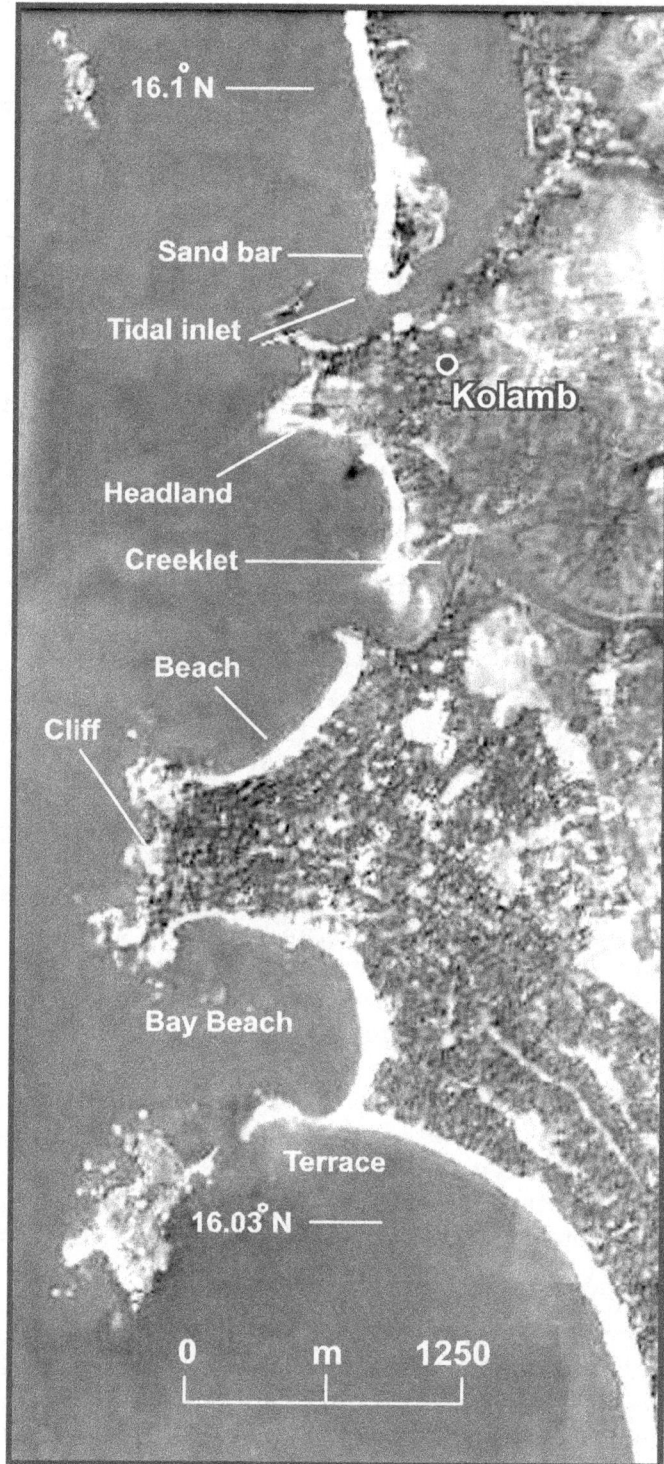

Fig. 29 : Coastal Features

The Beaches of Konkan :

Konkan is dotted with innumerable, small, sandy pocket beaches. Table 20 gives a summary of geomorphic aspects of a few representative beaches on this beautiful coast. Sandy beaches predominate the shoreline, but there are few mud beaches such as at Rewas and shingle beaches like that at Shekhadi. The sediment characteristics and the morphodynamics of these beaches are controlled mainly by specific wave and tide environment related to seasons and tidal range (Karlekar 2014). Wide beaches with a well-developed berm and beach face are characteristic of fair weather period.

Table 20 : Beach Characteristics Along Maharashtra Coast

Location	Lat/Long	TR,S	TR,N	Br. Dist.	Br. Length, km	surf zone W	swash zone W	Beach W, m	slope deg
Bordi	20,7/72,44	4.2	3.4	3.5	117	21	75.6	21.5	3.22
Dahanu	19,58/72,43	4.1	1.7	5.3	45.8	3.9	25.07	30.3	0.48
Wadhavan	19,58/72,40	3.9	1.6	1.4	132.5	31.6	93.5	8.9	4
Tarapur	19,54/72,42	3.9	1.6	5.9	355.8	100.5	167.9	95.5	1.47
Dandi	19,48/72,42	3.7	1.5	3.5	188.3	64.8	145.5	10.5	3.01
Satpati	19,43/72,42	3.6	2.6	4.8	125.3	43.9	95.5	20.3	0.3
Arnala	19,27/72,45	3.9	2	0.8	120.8	47.9	36.3	37.9	2.94
Mumbai	18,55/72,50	3.6	1.9	5.1	103.4	10.7	39.1	35.6	0.33
Karanja	18,55/72,56	3.6	2.1	1.5	72	20.1	32.4	34.4	2.64
Revas	18,49/72,57	3.6	2	6.9	222.3	46.3	84.6	66.5	0.86
Revdanda	18,33/72,56	2.5	1.8	4.1	239.6	58.9	100.5	106.8	1.55
Murud	18,19/72,58	2.5	1.8	2.1	226.6	66.1	129.5	72.9	0.55
s'vardhan	18,03/73,01	2.5	1.8	3.5	169.3	28.1	70.3	47.8	0.71
Bankot	17,58/73,03	2.5	1.8	1.3	937.8	273.9	492.2	121.7	5.1
Kelshi	17,55/73,03	2.4	1.8	2.5	148.8	51.8	40.4	14.4	7.6
Anjarle	17,51/73,05	2.4	1.7	2	114.1	56.5	21.02	41.7	5.7
Kolthare	17,39/73,08	2.2	1.6	1.5	51	15.3	13.6	20.4	8.5

		TR	S/N	Br					
Dabhol	17,35/73,11	2.1	1.5	0.9	483.6	84.4	299	85.3	9.7
Guhagar	17,30/73,11	**2.6**	1.5	6	77.8	33.4	22.2	18.2	6.3
Jaygad	17,18/73,14	**2.7**	1.6	2.4	105.6	27.1	32	35.2	10.2
Ratnagiri	16,59/73,18	**2.7**	1.5	0.8	85.6	45.2	35.7	28.2	1.1
Vijaydurg	16,33/73,20	1.8	1.4	0.6	98.1	32.7	43.3	41.6	2.8
Devgad	16,23/73,23	1.8	1.3	0.5	87.8	26.5	50.4	50	1.3
Malvan	16,03/73,28	1.9	1.3	1.7	159.8	35.1	23.5	64.5	4.9
Dabholi	15,53/73,36	1.5	1.3	2.3	151.7	34.3	15.45	48.6	12.1
Vengurla	15,51/73,37	1.5	1.2	5	102.6	22.4	24.74	66.4	3.2
Navabag	15,50/73,38	1.4	1.1	5.1	89.3	26.4	15.6	37	3.2
Tank	15,48/73,39	1.4	1.1	1.5	278.8	32.9	51.27	71.6	1.2
Shiroda	15,46/73,39	1.6	1.2	4.3	104.6	40	25.62	13.8	2.5
Average		2.6	1.6	2.9	179.1	47.64	79.3	46.4	3.7
Std Dev		0.9	0.4	1.8	173.8	48.4	99.9	29.5	3.2

(TR- Tidal Range, S - Spring, N - Neap, Br - Breaker)

Source : Author

They are transformed to narrow beaches with steep to very steep beach faces in monsoon with varying degree of steepness and beach cutting.

There is a considerable amount of variability in sandy beaches, which is a result of wave environment. The entire beach zone consists of depositional facies formed by waves; wave induced currents and associated flows.

Most beach sediments are well sorted and major differences in grain size reflect differences in wave energy levels. Tides are the main force in macro and meso tidal environment, in north Konkan. Decrease in the velocity of tidal currents at ebb, results in the sediment deposition in swash zone. The flood tide currents, on the contrary, induce erosion and cutting of beach profiles. In addition to these daily changes, Konkan beaches also undergo periodic changes related to seasons. Low, flat, swell waves during fair weather build up the berm or beach face and high, steep, storm waves in monsoon cut the beach face (Karlekar, 1997). Flat beaches in Konkan are usually associated with low and spilling breakers of fair weather whereas plunging breakers front steep beaches.

Swash aligned beaches (Davies, 1977) on this coast are found along indented and irregular stretches. In some cases they are transformed to drift aligned beaches in monsoon.

The ridge and the runnel and the rhythmic forms such as cusps, ripple marks, mega ripples, crescentric bars, berm and dunes are the essential morphological features seen on Konkan beaches. A great variation in size, shape and location of these features on Konkan beaches is very remarkable.

On fine sandy beaches the quantum of swash and backwash is more or less equal due to restricted percolation. However, on the coarse sandy and gravel beaches the percolation of swash waves is more effective. This causes vertical building and formation of steep beaches. Such beaches, however, are not frequent on this coast. They can be seen at places like Shekhadi, Uran and Karanja. The sandy beaches are steep only in monsoon when the maximum slope attained by beach is as high as 7 to 11 degrees. The average beach slope in fair weather is less than 2 to 3 degrees. In monsoon the sands are poorly sorted and show a positively skewed, leptokurtic distribution.

The sediments on Konkan beaches are subjected to reworking every year by aeolian, biological and coastal processes. Many beaches such as Kihim, Revas, Adhe, Anjarle, show a vertical sequence of fine sand, coarse sand, silt clay and mud even upto a depth of 40 cm. The subsurface sands are poorly to moderately sorted. The mud at depth is scoured, reworked and spreads on the beaches in monsoon (Kale and Awasthi, 1993). This phenomenon is especially seen at Avas, Revadanda, Kelshi and Adhe. This seems to be a recent event (Karlekar and Devane, 1995) and is restricted to meso tidal beaches.

Konkan is also endowed with long, beautiful sandbars and spits, which are essentially the sandy beaches attached to the main land at one end. Dandy, Revas, Revdanda, Devbag and Ubhadanda are some of the striking examples of sandbars on Konkan coast. They are usually drifts aligned, in that they are built parallel to the line of maximum drift. The building of beach abruptly ends where coastline followed by littoral currents, turns landward, at the entrance of tidal inlets. Spits like the one at Rewas, produced by the combination of drift and tide invariably show features of tidal and drift dynamics,as the bed forms. The swash-aligned beaches on Konkan coast are crescentric beaches. A few spits end in one or more hooks or recurves producing distal convexity.

Beach pits, low tide fans and mud balls are other important sedimentary structures characteristic of Konkan beaches. The occurrence of mud on beaches is a recent phenomenon and probably indicates a slight rise in sea level on this coast.

The cross shore and long shore dimensions of beaches on Maharashtra coast vary significantly. They range from narrow steep beaches to wide dissipative beaches all along the coastal stretch. These beaches also range considerably in length. At the longer and larger scale these beaches on Maharashtra coast include extensive beach systems with beach ridges, multiple bars, fore dunes and transgressive dunes. The near shore zone is a wave shoaling zone where decreasing depth results in increasing wave interaction with seabed (Short, 2000).

The Surf Zone : It extends from wave breaking point to the point where wave collapses and becomes swash. The wave breaking point ranges in distance several hundreds of meters. On this coast average wave breaking distance in fair weather is 179 m although it varies from 46 m at Dahanu to 938 m at Bankot. It appears that the surf zone width on this coast is not influenced by tidal range. The multibar and dissipative beaches such as Bankot, Dabhol and Tarapur however show that their morphology is controlled by shoreward moving decaying gravity waves from wide surf zones (Karlekar2014).

The Swash Zone : It is a relatively narrow zone extending from surf zone limit to upper limit of swash action. Swash follows a backwash and is thus a bidirectional flow. It is reported to be influenced by tidal range, beach slope and width. Increase in spring tide range on this coast produces wider swash zone. The swash zones near beaches on coastal sector from Dahanu to Revas where spring tidal range is more than 3.6 m are distinctly wider than those to the south. The width of the swash zone also increases as the increase in beach width.

Tidal Range : The tides on the coast are a non essential though ubiquitous component of many beaches. They however do contribute substantially to beach morphology. Usually wider, low gradient, featureless beaches are normally produced in the areas of high tidal range (Short 2000). On this coast however increase in spring and neap tide range shows reduction in beach width in general. Beaches at Kelshi, Kolthare, Guhagar,Ratnagiri and Shiroda are reduced in width considerably at spring tide however those at Revdanda, Bankot, Dabhol and Murud are wider than expected. At Spring as well as at Neap beaches at Wadhavan, Dandi, Satpati, Kelshi, Kolthare and Shiroda appear very narrow considering the level of tidal water. Beaches on the middle konkan coast from Bankot to Jaygad where spring tide range is between 2 and 3m are steeper than expected and on the south coast where tidal range is less than 2 m are gentler than expected.

Beach Slope: The beach width and its gradient are the most useful beach components which can be used in their classification (Fig.30). Narrow and steep beaches are usually classified as Reflective beaches (Davies 1977, Pethick 1984, Woodroffe 2002). These types of beaches are found normally to the south of Kelshi where tidal range is less than 2.4 m.

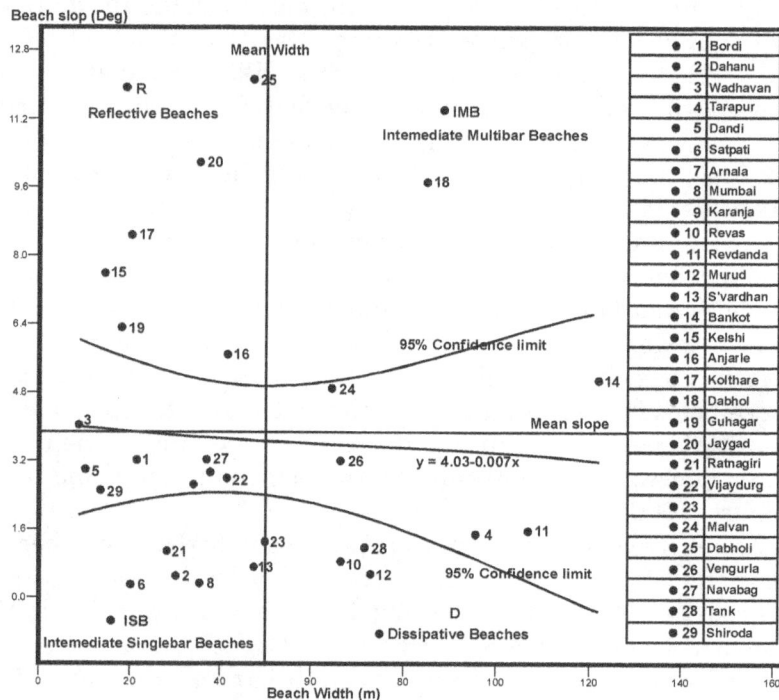

Fig. 30 : Classification of Beaches (Konkan Coast)

This however is not a rule and reflective beaches could be also seen where tidal range is around 3.9m (at Wadhavan). Similarly Dissipative wide, low gradient beaches are found at number of places although their significant occurrence is where tidal range is more than 2.5 m. Intermediate single bar beaches with less than 50 m width and less than 3.5 degrees slope and intermediate multibar beaches with more than 50 m width and more than 3.5 degrees slope are found at many places on this coast.

Beach Blisters on Konkan Coast :

Beach Blister is an important rhythmic micro feature of sandy beaches, variously described as sand dome, sand hole, air hole, beach pit or beach blister (Davis, 1978). There is very scanty information available on this feature of sandy beaches. The feature temporarily preserved on sandy beaches, is produced due to structural and textural inhomogenities of beach sediments and the intergranular space that traps the air and gives it out with tides.

The sea beaches on Konkan coast are dynamic systems which constantly adjust to fluctuations in local energy levels. This applies to all kinds of beaches, irrespective of the apparent uniformity or heterogeneity of the substrates. These adjustments take place locally and leave their imprint for a significant span of time on the beaches.

A detailed analysis of these sedimentary structures on the beaches has revealed an enormous variability; both perpendicular as well as parallel to the shoreline. This variability

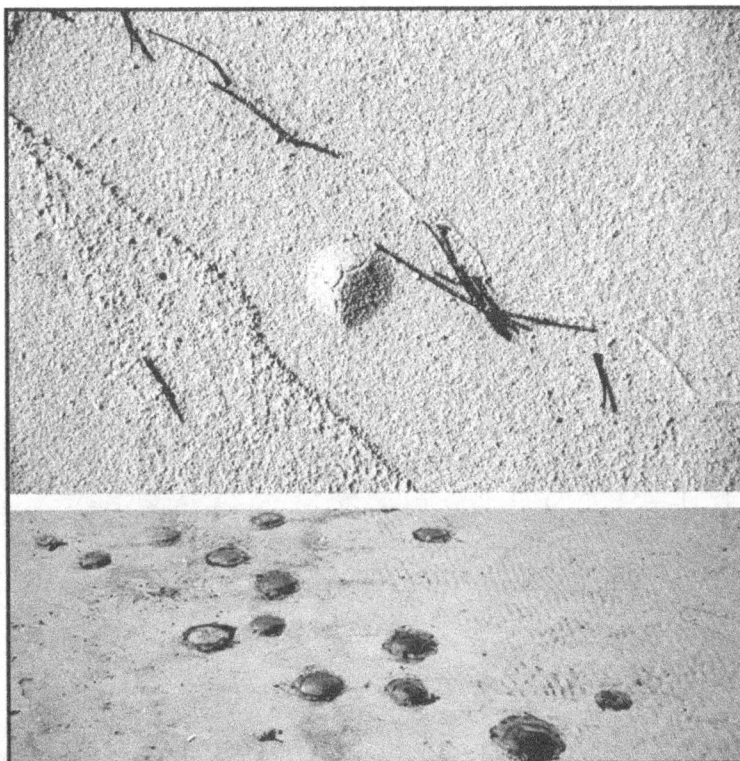

Fig. 31 : Beach Blisters

is still not properly understood.

The beach blisters on Konkon coast occur in a specific, narrow, elongated zone roughly parallel to the beach. They occur in the area of beach face or berm. They are seen parallel to each other and separated by few meters distance.

They are developed on fine sandy areas of the beach and are remarkably absent from the coarse sandy areas of the beach face. Such blisters are frequently observed on the sand beaches of south konkan especially on Guhagar, Velneshwar and jaygad beaches. Their occurrence goes on decreasing in down beach direction. It increases towards up beach direction. They occur as small, microscopic depressions covered by a thin cover of silt and clay forming a dome. Slight disturbance causes the domal bulge to collapse. The depressions exhibit a floor of silty sediments.

The blisters on the upper beach areas are relatively more firm and show signs of slight compaction. They appear as dark black spots on a plain white sandy beach.

The Coastal Dunes :

This is a well-marked and distinct feature of this coast. On the backside of many beaches primary dunes with characteristic wind ripples and parallel ridges of secondary dunes can be easily recognized. The embryo dunes, foredunes and backdunes are relatively

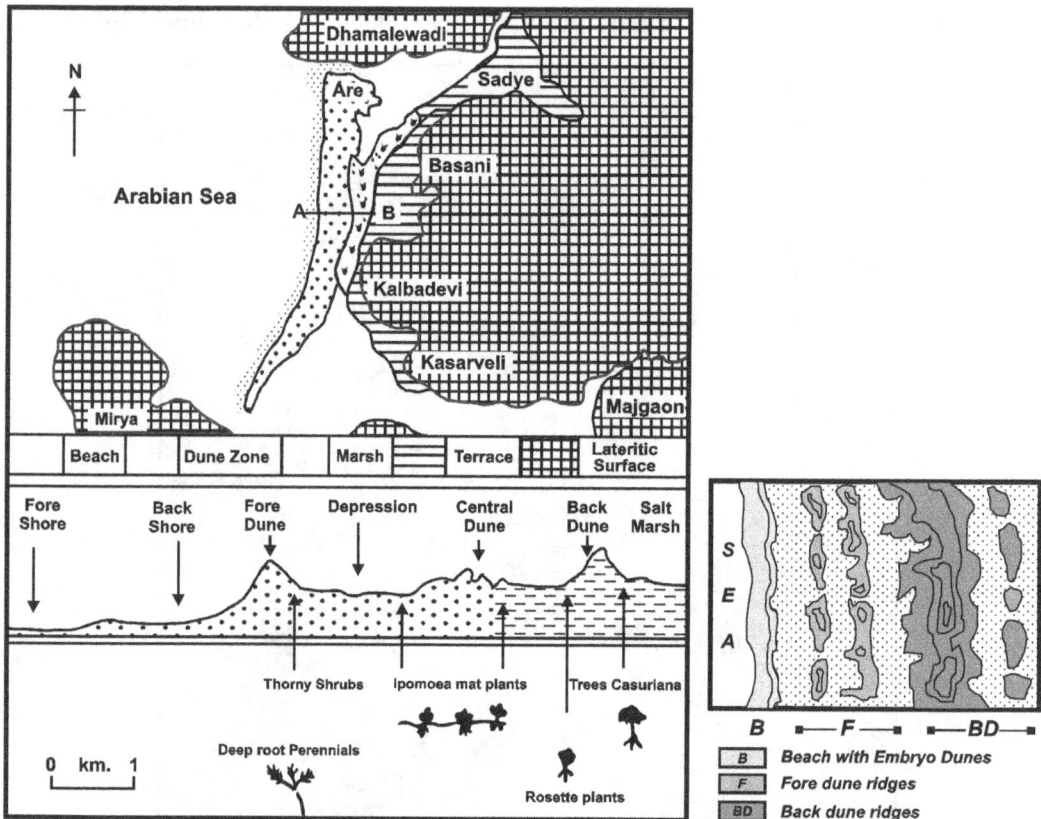

Fig. 32 : Coastal Dunes and Dune Plants

more conspicuous at Diveagar, Kelshi, Tambaldeg, Mochemad and Velaghar. There is a great variability as regards their morphology, orientation and the degree of preservation (Deswandikar and Karlekar, 1996).

On the narrow sandy beaches with a width not more than 50 m, the dunes are low and quite inconspicuous. The dune zone assumes certain amount of significance only on wide sandy beaches where they form extensive dune systems. The dune systems at Kalbadevi, Diveagar, Kelshi, Mirya, Sagartirth, Ubhadanda, Mochemad and Velaghar are typical examples of such formations. (Fig 32).The foredunes (outer and seaward) and the backdunes (inner and landward) are relatively higher than the central dunes. The foredunes actually form a dune wall sloping steeply seawards and gently landwards. The backdunes are more or less symmetrical but scattered. The scattered dunes give an impression that they are the residuals of ancient, continuous, fore dune ridge. The inter dunal flat between foredunes and backdunes is occupied by low, discontinuous, shapeless and scattered mounds of blown sand. The foredunes and to some extent central dunes are normally covered by dune plants like Ipomoea creepers, which form a thin mat on the dune surface. These plants have helped in trapping the sand blowing from beaches and thus building the dunes.

The most landward backdune ridge is usually a lithified, aeolian dune. Some of the important coastal settlements such as Nandgaon, Diveagar, Guhagar, Kolthare,Mirya are situated virtually on top of such lithified fossil dunes comprising of aeolinites locally known as Karal rocks (Dikshit, 1975, Deswandikar and Karlekar, 1996, Karlekar and Gadkari, 1998,). (Fig. 33)These are the places of ample and shallow groundwater that is utilized by

Fig. 33 : Village Kolthare and Nandgaon Situated on Old Fossil Dune

the local people for drinking as well as growing coconut trees.

All dunes are covered by dune plants which either form a thin sheet like a mat or stand in groups forming patches. What is important is that the plants all over the dune zone are not alike and show a distinct type variation from the foredunes to the backdunes.

The interdunal flat is dotted with a number of small circular depressions where thin grass grows. Considering the height, the density and the type of plants growing on the dunes, the foredunes can be classified as the primary dunes. The accretion of the sand and the building of these dunes are impeded due to the thorny shrubs which are poor sand trappers. Although the foredunes form a wall like barrier along the beach they seem to be basically nucleus oriented. That is to say that their growth is favored at a certain point where there is a sudden drop in the velocity or where there are small thorny shrubs initiating the accretion of the sand particles. It appears that the latter condition is responsible for the building of these foredunes, as the windblown sand can easily reach further inland up to the central dunes.

On the interdunal flats, less stable, drier windblown sand has accumulated and given rise to low dunes. Generally these dunes are perpendicular to the foredune ridge. The low central dunes are covered under a thin mat of 1pomoea plants.

The backdunes are also relatively prominent but are eroded and discontinuous. Here the sand is less mobile and readily settles down. In fact this might be the older foredone ridge when the sea level was slightly higher than the present one. One can classify these dunes as the secondary ones which are either lithified or stabilized. The most significant aspect of these dunes is their internal horizontal bedding. Back dunes are the product of ancient a aeolian processes and are retained in the landscape since the last fall of sea level. They are beyond the reach of present day wind and waves. Moreover the sand layers in these dunes are relatively more cohesiveness and compact.

A 40 to 50 m wide zone bordering upper beach along its landward margin and sometimes stretching for about 300 to 350 m in the north south direction is characterized by number of shadow and embryo dunes. These sand mounds are not more than 1 m in height. Embryo dunes are covered with Ipomoea creeper and Sphinifex grass.

The Dune Plants :

The major aspect of the dune zone is the dune vegetation. Ipomoea creepers are the dominant plants which stretch over a wider dune zone. The Ipomoea creepers are usually thick and dense in the interdunal area. Elsewhere they are replaced by rosette plants, thorny shrubs or deep rooted perennials. Thorny shrubs grow on the leeward of the foredunes, the area being comparatively drier and sand grains less mobile. This type of a environment helps the plants sending their roots deep into the dunes. These shrubs and deep root perennials help build the dunes throughout the year. The plants are scattered and nowhere assume a dense aspect. The secondary back dunes are different from the others in that their basal parts are covered by rosette plants. These rosette plants are microscopic and do not help much in trapping the sand. It was also observed that these rosettes grow more in the drier season and diminish in number in wet monsoon period. The bigger and higher secondary dunes are characterized by plant cover. The casuarina trees occupy the leeward slopes and the trees with a distinct rhizome system occupy higher and windward slopes. These trees act as a very effective screen to the sand grains.

The Sea Cliffs, Shore and Rock Platforms :

Impressive sea cliffs, shore and rock platforms characterize the rocky coast of Konkan. These features are usually found along joints, cracks and other weaker sections of the rocky headlands.

The destructive impact of waves along this coast is often far greater than is generally realised (Karlekar, 1981,1993 a,2009). The headlands and promontories of the coast are subjected to shocks of enormous intensity especially in monsoons. Cracks and crevices are quickly opened up and extended. High-pressure spray of waves is forcibly driven in to every opening. At any given time and place the actual form of the sea cliff and caves depends on the nature and structure of rocks exposed and the relative rates of marine erosion and subaerial denudation.

Cliffing is the dominant process on this coast. The sea cliffs in the area south of Vasai (19.4^0 N / 72.8^0 E) usually show a wave cut bench at the foot. The predominant cliff face angle is found to be about 60 degrees. Average height of the cliffs is around 9 m. Overhangs and notches slightly above the influence of present day waves characterize the lower sections of the cliffs. cliffs and caves or overhangs are the main profile components seen at majority of places such as Korlai, Hareshwar, Kolthare, Velneshwar, Jaygad, Agargule, Vetye, Kunkeshwar, Sarjekot, Kondura and Boriya Bandar (Fig.34).

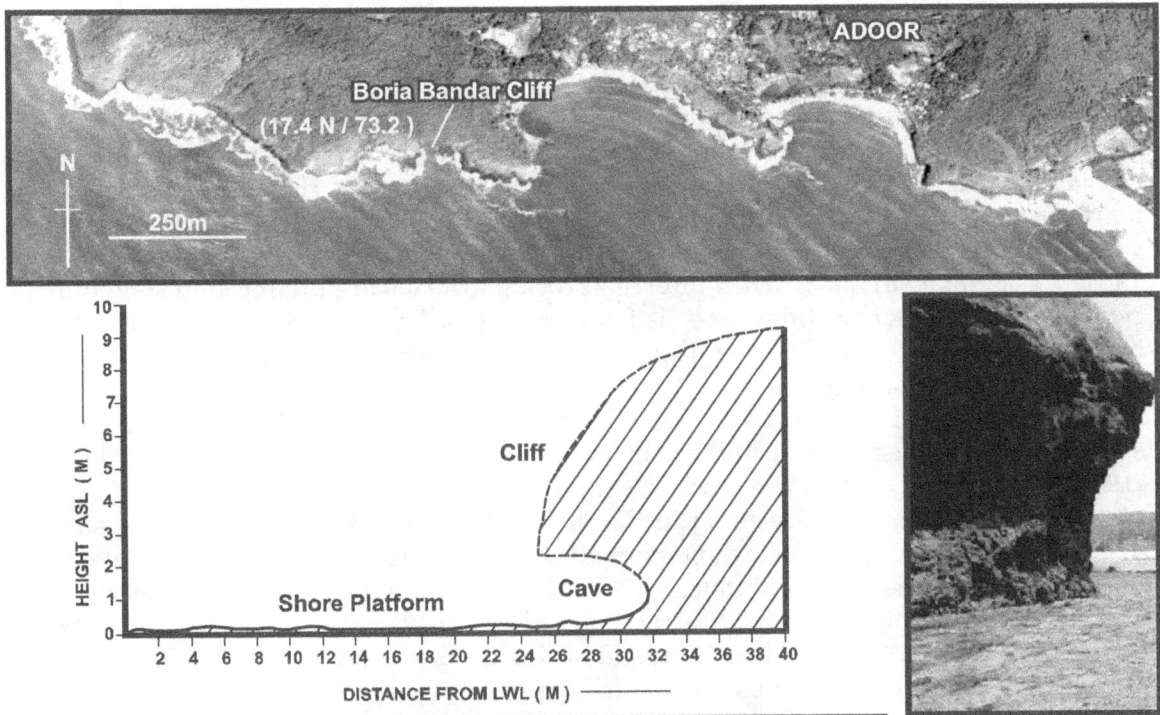

Fig. 34 : Sea Cliff Profile at Boriya Bandar

The near absence of quarrying material at the foot of the cliffs, insignificant mass movement on the forested upper sections and the lateritic cover on the cliff tops especially in middle Konkan suggest insignificant subaerial erosion. The spray marks, caving and undercutting however confirm strong marine erosion of the lower sections of the cliffs.

The form and nature of the sea cliffs characterized by overhangs and notches do suggest a slightly higher sea level in recent past. At number of places the overhangs and notches along with the honeycombs appear abandoned and incompletely developed.. The inland distance at which dead cliffs occur is around 350 to 500 m. which is also significant in this regard. At few places the distance inland is not very striking but the height above present sea level is important. Here old cliff-like faces modified by sub aerial processes appear roughly at a height of 9 to 10 m. above sea level.

The shore platforms at the foot of the cliffs are also a striking feature in Konkan. Their average width rarely exceeds 30 m. The platforms are intertidal and are shaped by abrasion and water layer weathering. A low tide cliff of about one meter height bordering the seaward margin is a typical feature of the platforms on the Raigad and Ratnagiri coasts. The surface of the platforms all along the coast is dotted with innumerable shallow pools and potholes of varying sizes. The shore platforms are mainly produced in basalts although there is large number of lateritic platforms also. Lateritic shore platforms are characteristic of places like Ambolgad, Devgad, Kunkeshwar and Bhogave.

The features like Geos and Blowholes, Sea stacks and stumps are not very common. They are seen only at Korlai, Velneshwar, Hedvi and Barshiv.

The Rock Platforms :

Rock platforms are conspicuous elements of rock coasts in many parts of world. They are made principally of hard and resistant rocks with minor softer formations. Wave erosion along horizontal to gently dipping bedding planes gives rise to the gently sloping platform surfaces. Erosion also breaks off rectangular blocks of rock along intersecting vertical joints. This controls the shape and orientation of the platform margins (Trenhaile 1987, Masselink *et al* 2003). Such intertidal rock platforms are produced in semi diurnal tidal environment where tidal range is more than 4 m. They may be separated from and backed by beaches. They are thus different from shore platforms which are developed at the foot of sea cliffs. (Fig. 35)

Along North Konkan Coast especially to the north of Vasai (19.4⁰ N / 72.8⁰ E) and especially on Palghar Coast of Maharashtra between Zai (20.13⁰ N /72.73⁰ E) and Tarapur (19.86⁰ N / 72.68⁰ E)and on the coast of Satpati and Vadarai there are very striking rock platforms with sharp surface relief in the inter-tidal zone. They are backed by silt clay deposits, sandy beaches, dunes, dune systems and wide littoral terraces slightly above the high tide level (Karlekar and Shitole,2013).Rock platforms with a slightly steeper gradient and fronted by beaches or littoral terraces are also observed along Raigad coast between Borlai (18.5⁰ N / 72.9⁰ E) and Northern end of Kashid Beach (19.4⁰ N / 72.9⁰ E).

All these platforms in question are developed at the level of low tide. The platforms are 'non structural' benches, which cut across the local structure of coastal rock formation at the height of 2-3 m above mean sea level.

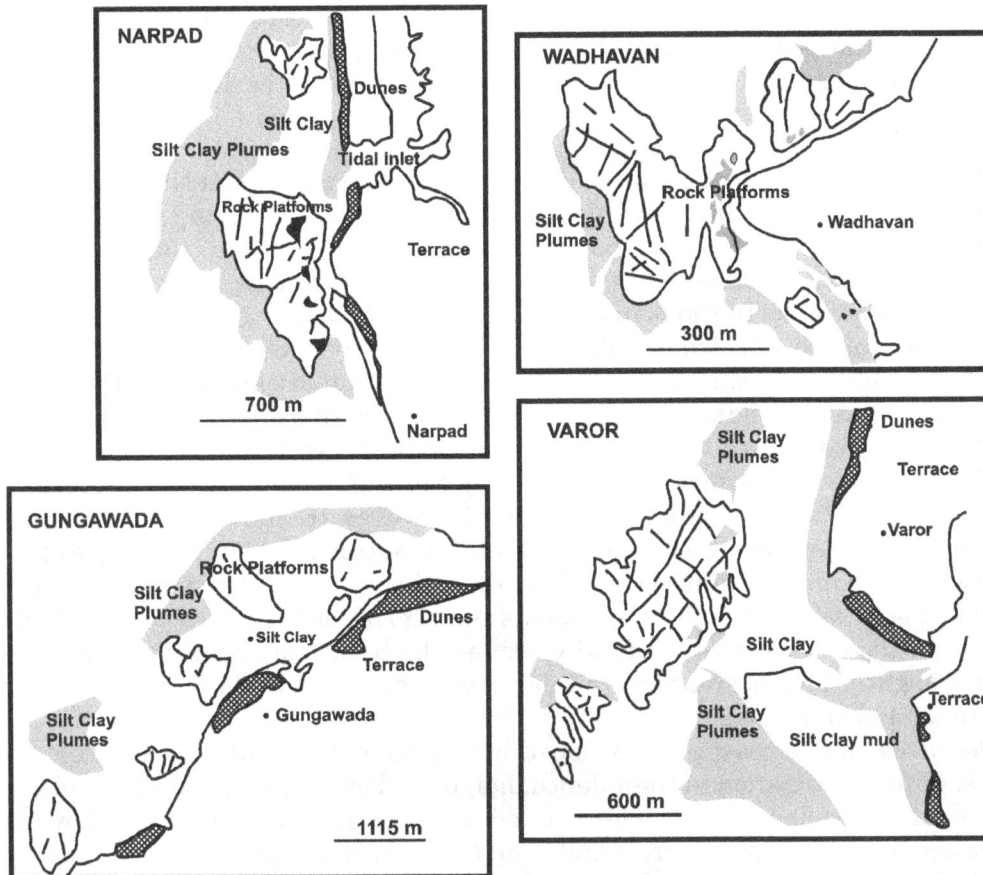

Fig. 35 : Rock Platforms

The Features of Rock Platforms on This Coast can be Summarized as :

1. They have a width that ranges between 800 and 2450 m
2. The surface of the platforms is leveled in intertidal zone and especially between the mean sea level and mean high tide level, except seaward margins
3. They are developed on basalts with strike nearly parallel to the general direction of the shoreline and dip of 1in 250 toward sea
4. Their surfaces show very sharp relief lines, furrows and pools
5. The surface inclines slightly towards sea with an inclination of 0.06° to 0.83° especially in the middle sectors of the platforms.
6. Along the rim of a few platforms there are slight breakwater like ramparts
7. On the surface there are frequent rounded gravels and micro regions of silt sand deposits.
8. All the platforms have been carved with deep wave furrows
9. There are no cliffs at the rear of the platforms. They are separated from but backed by beaches.
10. On the surface honeycomb structures, solution pools, potholes filled with

sediments and mangrove bushes on silt clay deposits are found.

11.   The platforms are carved by wave furrows which are regulated with joints.

12.   They are produced by a process of water layer weathering.

The Sea Caves :

The sea caves represent a link in the evolution of sea cliffs and shore platforms. The caves at various elevations above present sea level, their varying sizes, shapes and orientation, sedimentary material on the floors of the caves, their relationship with cliff sections and shore platforms at the foot, are very significant factors since they throw a light on the sea level fluctuations in the area (Fig. 34).Active caves on Maharashtra coast, as elsewhere, are those which are shaped by present day waves and tides. They are the result of undermining by waves. Abandoned are those that are away from the influence of present day waves and slightly at a higher elevation. On this coast they are many a times characterised by coarse gravel deposits at the foot and vegetation growth. Through the base of active cliff is continuously attacked by the waves, the process of erosion is not effective in fair weather season because then the waves are moderate. During monsoon season as the wave energy is more, erosional activity is stronger. Most of the caves developed in Konkan, face South West direction. The predominance of cliffs and caves on the South-West facing slope of the headlands suggests their formation due to south westerly wave attack in monsoon. Sea caves generally occupy the basal portion of the sea cliff. Older caves are slightly higher and farther inland, whereas present day caves are lower and nearer to the low water line.

The caves on this coast are developed in basalts, granites as well as laterites. They show a significant variation in their depth, height and overall form. Most of the caves are typical examples of abandoned caves fronted by shore platforms. Caves at few places in the area are fronted by long narrow and deep Geos which are produced along the weaker zones. Caves found at some height are relatively narrow and small and insignificant as compared to those developed near the foot of the cliff. Ancient caves and cliffs are located 20 to 30 m. inland from the low water line. Caves at Harihareshwar and Are near Ratnagiri show honeycomb structures produced due to wave action.

Caves on Konkan coast is the most striking evidence of undermining by sea waves. They are excavated along belts of weakness of all kinds and especially where the rocks are strongly jointed. By subsequent falling of the roof and removal of the debris, long narrow rock inlets are developed. The tidal rock inlet of this kind is known as `Geo'. Such Geo features can be seen only at few places like Hedavi and korlai.The roof of the cave at landward end is sometimes found may communicating with the upper surface by way of vertical shaft. A natural chimney of this kind is known as a 'Blow-hole'. The name blow hole refers to the fact that during monsoon, spray is forcibly blown into the air each time a breaker surges through the cave. Hydraulic action of the wave compressed air is responsible for the development of such blow holes.They are not common and are seen only at few places like Ratnagiri.

Shoreline / Littoral Terraces :

Narrow, flat terraces are seen in backshore areas all along the coastline. These terraces are small, elongated and usually parallel to the coast. Their height varies from 3 m to 7 m above sea level. They either stretch between two streams to north and south or are confined to regions surrounded by hill slopes (Fig. 36). Their orientation in most of the cases is north south..The water table on the terrace is relatively shallow and helps vegetable farming. Sometimes they are fronted by sandy beach or even a lithified old sand dune.

Fig. 36 : Shoreline Terrace at Wayangani

The morphology, shape and the configuration of these terraces undoubtedly points to their marine origin and defunct nature due to a slight drop in sea level. Steep banks of the channels of small streams, which cut across the terraces, suggest an increase in fluvial activity, subsequently after these terraces were developed. The lower reaches of the streams on the terraces very near the shore are necessarily tidal and can be attributed to a very recent, slight rise in sea level. A sharp and abrupt contact of terraces with the inland hill slope edges is also very striking.

The Tidal inlets :
Estuaries and Creeks :

The estuaries and creeks on this coast are distinct especially due to their tidal and fresh water regime. They also exhibit a complex pattern of sediment input.

The tidal inlets to the south of 18 degrees north parallel are wave dominated. Northern estuaries have a strong tidal control. Most of the estuaries on Konkan coast are NW – SE oriented and suggest a structural control in the tidal sectors of the streams. In north Konkan, creeks and estuaries are found to be bar built and coastal plain estuaries. Lengthening of ebb conditions is an important aspect and is reflected in the tidal delay period (residence time) of about 1 to 2.5 hours (Karlekar, 1996). Imbalance between the length of the estuary and the contemporary tidal range is seen in the ponding of tidal water in mid portion of some estuaries like Kelshi and Anjarle.

The major sedimentary environments of Konkan creeks and estuaries include high and low tide sub tidal and intertidal flats, sand banks mangrove swamps or salt marshes and scoured tidal channels (Karlekar and Keskar, 1993). (Fig.37 & 38).They are produced by site specific hydrodynamic conditions like wave action, flow velocities, turbulence, mixing and scouring. The mid estuarine sectors are invariably the areas of silty clayey bars. On an average the depth of these tidal inlets varies from 1 m near the head to about 4 m near the tidal mouth. Tidal water penetrates to a distance of more than 30 m in many estuaries like Amba, Kundalika, Dabhol, Arjuna and Karli. The estuaries on middle and south Konkan coast are narrow, elongated inlets with relatively little human interference. The deposition of placer minerals is also a characteristic of a few tidal inlets (Karlekar, 2001).

Fresh water flow in monsoons is one of the fundamental controls and affects salinity structure of these estuaries (Karlekar, 1996). A pronounced salt wedge in dry season is a dominant feature. During monsoons due to strong fresh water flow, all traces of seawater are effectively flushed out from the headward sector of these inlets. However, some amount of vertical stratification still remains in the lower column of tidal water. In post monsoon the salt wedge is re-established very rapidly.

There exist several areas of salinity and sediment concentration in most of the Konkan estuaries. A substantial portion of suspended sediments that enters the estuaries is deposited within the estuaries only. A large proportion settles on the mud flats and other areas outside the main tidal channel. The deeper areas act as sediment traps. Sedimentation in these estuaries appears to be governed by a whole set of factors like length of the tidal inlet, tidal range and the process of flocculation.

Following are the Salient Trends in the Hydrodynamics and Sedimentation of Konkan Estuaries.

1. Roughly to the south of 18° N parallel, the estuaries are wave dominated. To the north they are tide dominated.
2. The average tidal range increases from South to North and ranges between 1.2 m to 5.1 m.
3. Most of the estuaries are oriented NW-SE.
4. The estuaries in the North Konkan are bar built and coastal plain estuaries.
5. On an average, the depth varies from 1 m to 4 m, and length varies between 5

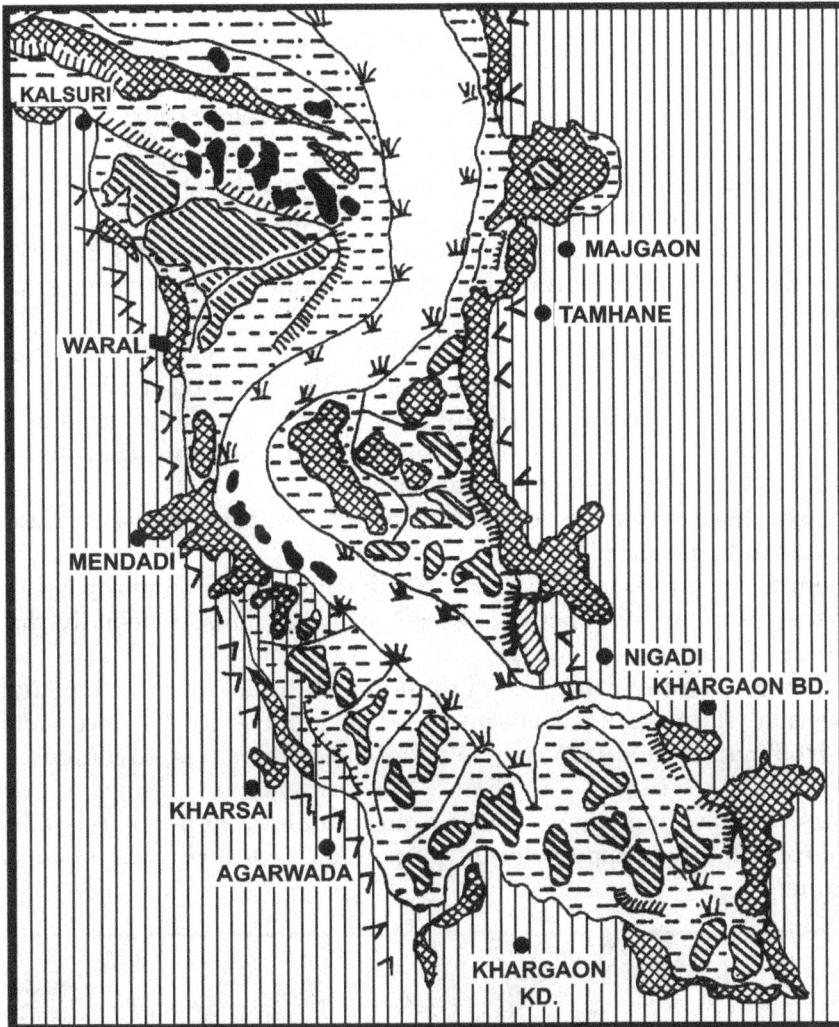

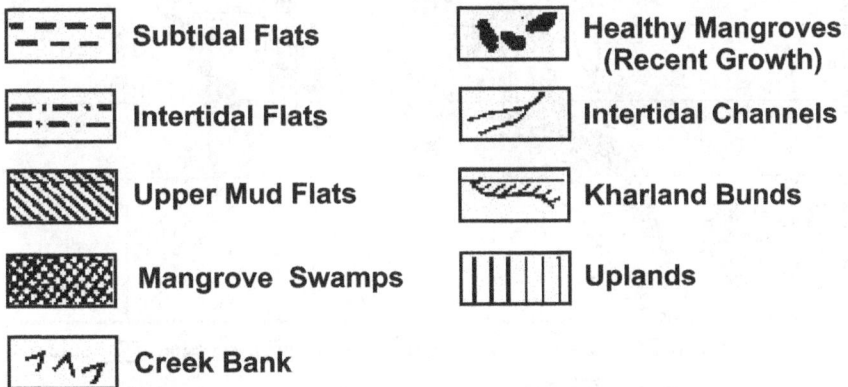

Fig. 37 : Sedimentary Environments of Mhasala Creek

Km to 10 Km.

6. lengthening of ebb conditions is an important aspect of these estuaries. The period of tidal delay (Residence time) is around 1 to 2-1/2 Hrs.

7. All the estuaries are seasonal in nature and therefore there is a marked seasonal shift of salt wedge.

8. Big estuaries show well mixing. Small are partially mixed.

9. Salt water/ fresh water ratio keeps on varying.

10. Most of the southern estuaries show a ponding effect in mid portion.

11. The organic matter content increases towards marshes/swaps and towards the mouth.

12. The material in the suspension is mainly derived from estuarine margins. Ebbing tides bring the material to lower zones. The flooding induces ponding, which redistributes the material throughout the cross sectional area of estuary.

13. The northern estuaries show a high rate of sedimentation due to industrial pollutants, and metallic flocculation.

14. The southern estuaries are narrow, elongated, tidal inlets with non significant human interference.

15. The mid estuaries are the areas of silly/clayey bars in many cases.

16. There are many evidences of sea level fluctuations, especially the higher still stands of Holocene.

17. The major sub environments of Konkan estuaries (Marsh edge, high tidal flats, low tidal flats, scoured channels, sand banks) are the products of specific hydrodynamic conditions (wave action, flow velocities, turbulence, mixing and scouring).

18. Marsh and swamp edges are the areas of dominant deposition of silt and clay.

19. Scoured channels especially in the northern estuaries show heavy deposition of mud.

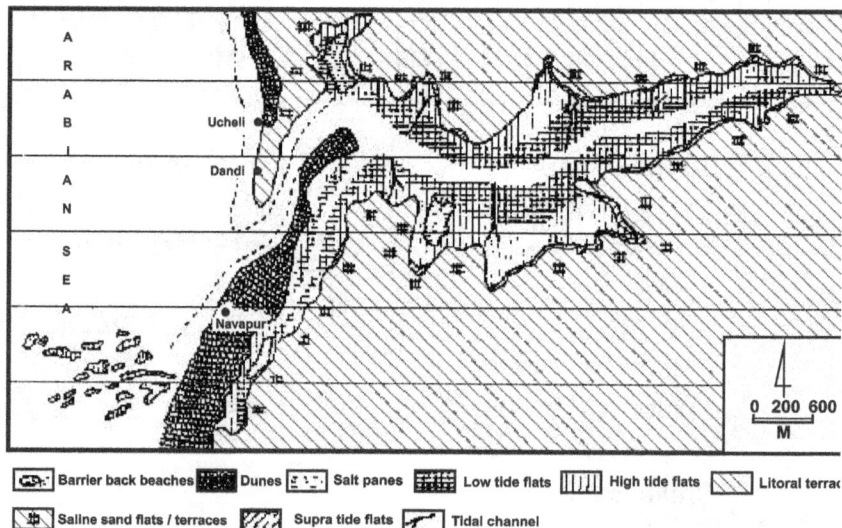

Fig. 38 : Sedimentary Environment of Ucheli Creek

The upper limit of the konkan estuaries is determined not by salinity but by the tidal ingress. This is because the steepness of the salinity gradient and the distance of marine water intrusion depend upon the pressure of the tides entering the estuary. Thus the upper limit of the estuary is determined hydrodynamically. The lower limit is fixed by geomorphologic features like rocky shore, delta region or tidal marsh zone. The variation in the size and shape, mixing characteristics and biocenosis of the estuaries is a result of the form of the original inlet, supply of riverine sediments, and marine sediments. The Konkan estuaries could be classified as sharply stratified (salt wedge), partially stratified (with significant vertical density gradient) and well mixed estuaries. Geomorphologically they may be classified as coastal plain estuaries, fiord type estuaries and bar built estuaries. Every estuary has its own personality described by distribution of sand silt bars, deep channels, shallow bays, tributary rivers, islands, and manmade jetties and harbours.

The Tidal Mud Flats :

Saline intertidal mud flats within tidal inlets especially the estuaries are the prime areas of sedimentation on Konkan coast. The Tidal mud Flat deposits on the ancient high tidal flats are covered under coastal alluvium. In the intertidal zone thick tidal mud is found on the surface, especially along the present high water line. During ebb, an extensive area of intertidal zone is exposed and one can see mud, sand and sandy mud everywhere in the inlet regions (Karlekar. 1993). The subsurface stratigraphy of the ancient mud flats reconstructed from bore hole data at Bokadvira (18.9⁰ N / 72.9⁰ E) shows a surface cover of coastal alluvium that ranges in thickness from 2.8 m to 9m (Karlekar 1996).This is underlain by a thick layer of soft marine clay 5 to 18 m thick (Fig. 39). This is a major subsurface stratum. The deposits consist mainly of soft clay, silt and

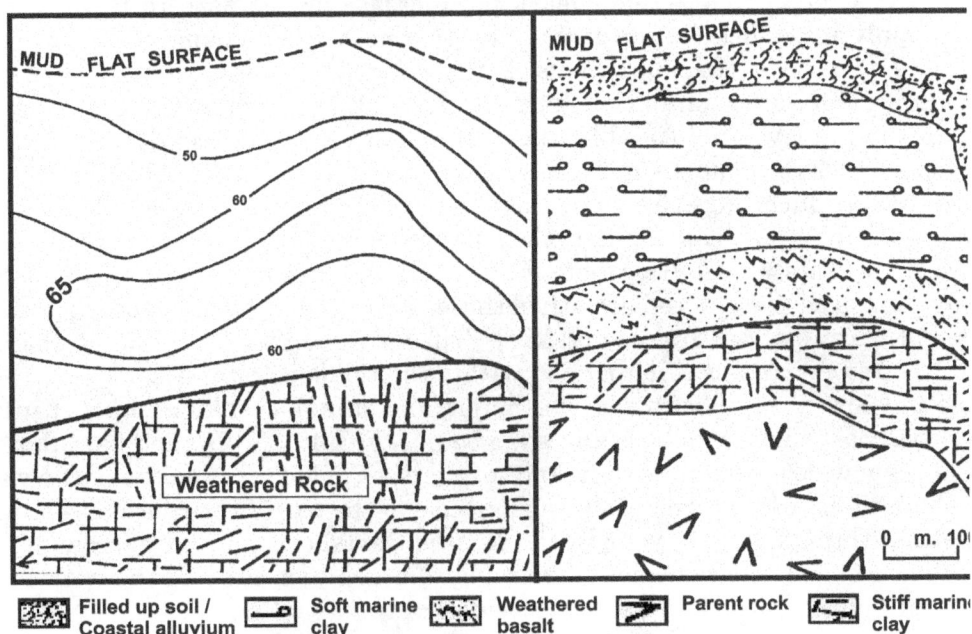

Fig. 39 : Tidal Mud Flat Deposits

small shell fragments of gastropods and plant detritus are also found at some locations like Bokadavira near Uran. More clay in the subsurface sediments is found at a depth of 8 to 15 meters and forms an elongated wedge of mud deposits at this depth. The clays in this subsurface layer are inorganic clays of high plasticity; inorganic clays still deeper have a moderate plasticity. The surface layers of coastal alluvium also contain some clay, but here well graded sand is the most dominant component of sediments. Well graded sand with small clay content is also found at 15 m. depth in some areas. A narrow region of stiff marine clay is found below soft marine clay in some sections. One can get a fairly good idea about the surface and subsurface stratigraphy of tidal mud flats in Konkan from the occurrence at Bokadvira described above.

Sand Bars and Sand Accumulation Features :

The sand accumulation forms, often-called sand bars or sand lenses are an important geomorphic feature of modern tidal inlets of konkan coast. These moderate to large size bodies of accumulated sediment are even found just outside the river mouths where they are submerged. Supply of fluvial sediment throughout the year and its non-removal by the waves and tides from the tidal inlet are most important factors that determine the development of these accumulation forms in the lower tidal reaches and the entrance of tidal mouths. The factors such as sea waves, tidal currents, sea level fluctuations, river discharge, sediments and shoreline configuration determine their morphology and formation.

These sand bodies have started showing tendencies of shifting within the inlet since last decade or so. Reduction in the size, change in the shape of the accumulation form, linear growth in upstream and downstream direction and in few cases change in the type of sediment is some of their salient features. Increase or decrease in fresh water and sediment supply and redistribution of available sediment due to changing tidal circulation pattern has produced the spatial variation in morphology, sedimentology and configuration of these accumulation forms in the tidal stretches of rivers in konkan. Fluctuations in sea level, indicated by the change in tidal structure along the coast, are also responsible for the changes in these sedimentary forms.

At few places such ridges and bars have increased in height and are obstructing the inland navigation. At majority of places they are swept away by strong waves in monsoon. Sustained heavy rainfall, Considerable river discharge and huge amount of suspended sediments control the extent and growth of these ridges in Konkan.The bars appear to be a permanent feature at few places and keep drifting throughout the year in the limited tidal area. Narrow channels remain open within the tidal sector or creeks.

The outflow area of many of the major and minor creeks and tidal inlets on Maharashtra coast show such bars and sand bodies (Fig. 40). A big sand bar, about 2 km long and 1.3 km wide exists at the entrance of a creek 2.5 km away in the outflow area of river Vashishthi near Dabhol on Maharashtra coast. The bar is a permanent feature and keeps drifting throughout the year in the limited nearshore area. A narrow, 6 m deep channel remains open near the southern bank of creek just close to the entrance. The average depth of the bar is 3 m below the sea surface.

The seasonal closure of tidal inlets is a common coastal phenomenon especially along the south Konkan coast (Karlekar 2011). It usually starts at the beginning of fair weather

Fig. 40 : Sand Bars in Terekhol Creek

around November when the average river flow drops down significantly. Hereafter persistence in wave direction and wave conditions remain more or less uniform for several days. River generated flood currents decline at the end of rainy season and wave energy is reinstated as the major sedimentary process.

The seasonal closure of tidal inlets is a common and important coastal phenomenon along the south Konkan coast of Maharashtra(Fig 41.). The closure of tidal inlets usually starts at the beginning of the fair weather or dry season around November when the average river flow drops down significantly. Hereafter persistence in wave direction and wave conditions remain more or less uniform for several days. River generated flood currents decline at the end of the rainy season and wave energy is reinstated as the dominant sedimentary process. Sediment from offshore sand bars are transported onshore and alongshore resulting in gradual narrowing and shallowing of tidal inlets.

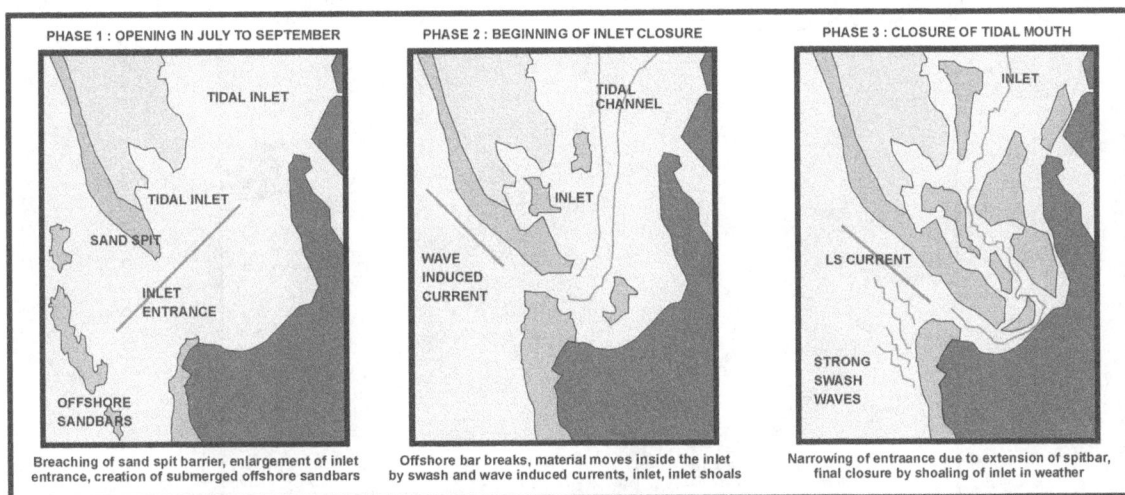

Fig. 41 : Seasonal Closure of Tidal Inlets

In all tidal inlets on this coast the existence and persistence of an entrance is dependent on the relative strength of tidal currents and river generated currents which maintain the inlet entrance as well as close the entrance by depositing marine sediment (Karlekar,2011). The balance between these processes determines the nature and the persistence of an inlet or a river mouth.

The Mangrove Swamps :

Mangroves are thickly vegetated intertidal estuarine wetlands, confined to silty bog formations on Konkan coast. They are restricted mainly to the sheltered shoreline zones regulated by tidal flooding.

The most widespread genera of mangroves on this coast are Rhizophora (Red Mangroves) and Avicennia (Black Mangroves). All shrubs and trees in mangrove swamps are characterized by adaptations to loose, wet substratum, tidal submergence and changing salinities. Prop roots and aerial breathing roots are the most frequent adaptations. The flora and fauna of mangroves in Konkan is zone specific and governed by temperature, salinity and period of tidal exposure. Everywhere they are the nurseries for fishes and crustaceans.

They are usually found near muddy creek banks comprising of fine silt and clay, which are rich in organic matter content. Their dense growth is seen in the tidal sectors of Ulhas, Amba, Savitri, Mhasala, Kalbadebi, and Rajapur creeks (Fig. 42), where shores are free from strong waves and tidal velocities are not very high. Here, the vertical tidal range between 2 and 3 m and a wide horizontal range has favored the thick growth of these halophytes on the coast. Important varieties of mangroves on this coast are Rhizophora mucronata, Rhizophora apiculata, Brugiera symnorhiza, Avicennia officinalis and Lumnitzera racemosa (Deshmukh, 1989).

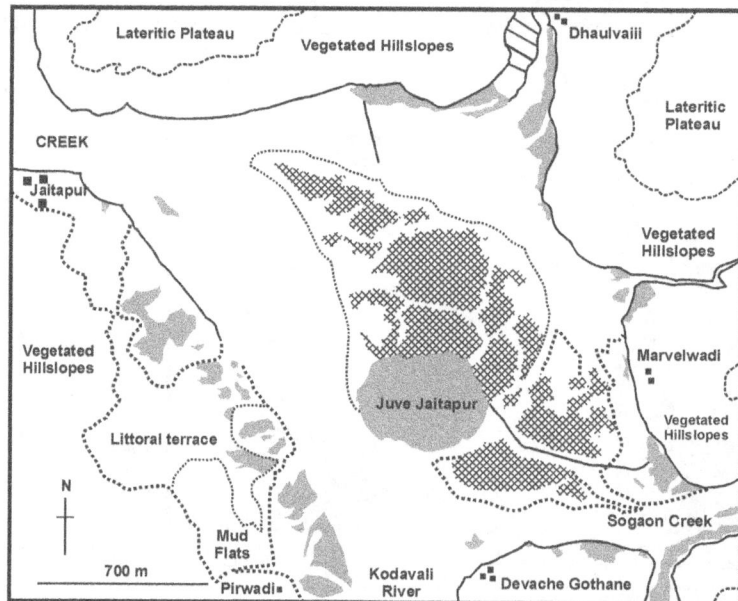

Fig. 42 : Spatial distribution of Mangrove Clusters (Numbered) in Jaitapur Creek

(C) Coast of Goa

Along the central west coast of India, the coast of Goa lies between 14°48' N and 15°48'N latitudes and 75°40'E and 74°20'E longitudes. It is considered to represent part of the Konkan coast. It is a narrow coastal strip located at the foot of the Western Ghats (Fig. 43). The foot hill areas are the plains at an elevation of about 100 m ASL. The height of western ghats in Goa state is nearly 1000 m. The foothill to sea region is 40 m km wide.This region is characterised by several small hills scattered all over the area.

Physiographically the region can be broadly classified into: 1) the coastal tract 2) sub-ghat region and 3) the high ranges of the Western Ghats. The drainage in the state is controlled structurally along E to W and NW or NNW directions.. Laterite covers the major portion of Goa and typically occurs as plateau laterite.It occurs at levels from 120 m to sea level. Remnants of the plateau surfaces showing deep lateritic weathering profiles are common in the northern part of Goa (Wagale, 1993). The southern part presents a rugged topography with hills having an altitude up to 600 m (Fig. 43).

Fig. 43 : State of Goa

The coast is prograding along beaches and retreating along the cliffs and headlands. Fluvial, marine and aeolian features are common along the coast. The presence of laterite beds along the estuaries at a depth of 27 to 35 m below present sea level indicates a drowned nature of earlier river valleys. According to Wagale (1993) the present coastline of Goa is neither emergent nor submergent.It is a combination of both. Occurrence of conglomerate beds, beach rock, beach ridges, old tidal flats beyond the present day high water line are the evidences of the sea level oscillations during the Quaternary.

The sediment transport on this coast is bi-directional. The overall net shore drift direction along Goa coast is towards south. It is qualitatively determined that the shore zone has long-term stability (Wagale, 1993). The coast all along its length experiences mix diurnal tides.

During southwest monsoon period, strong southerly currents erode protruded sectors and deposit eroded material, along varying sector. During rest of the year, under the influence of a northerly current, accretion is taking place along retreating sectors (Kunte et al, 2001). It was estimated by Antony (1976) that about 9×10^6 cubic m of littoral drift takes place along Calangute beach annually. It was also concluded that waves approaching from W and WSW generate inshore-offshore flows. The converging and diverging longshore flows, give rise to circulation cells in the nearshore regions. It was seen that the sediments on Colva beach mostly get recirculated between the two promontories, under the influence of the prevailing currents and the associated circulation pattern in the surf zone(Kunte et al,2001).

Goa coast is 135 km long rocky shoreline characterized by innumerable pocket beaches. The northern coast resembles more to South Maharashtra coastline. It is characterised by old and recent tidal flats, dune systems comprising of fore and back dunes, sea cliffs and shore platforms (Fig. 44). The beaches to the north reveal a retrograding shoreline where erosion is dominant.

Very narrow to wide pocket beaches vary in length from 400 m to slightly more than 3. 5 km and are backed by 6 to 8 m high, long and wide dune systems comprising of fore and back dunes (Table 21). The shoreline or littoral terraces on this coast are found at 6 to 9 m height above present sea level. Wide tidal flats having a width of about 350 m are seen within the estuaries like Terekhol and Chapora.Low sea cliffs with a height not exceeding 12 m above sea level are found along northernmost coast around Keri and Arambol only. The shore platforms at these places are 20 to 25 wide and characterised by furrows and pools.

North Coast : At Keri, the Coastal Area from Terekhol southwards is characterized by stable Sand Dunes in its pristine form. The fore and mid dunes are not very prominent. The back shore is well stabilized with a thick plantation of Casuarina.Along Arambol (15.7^0 N / 73.7^0 E) the sandy shore has very well developed fore and mid dunes and theyre are well covered with Spinifex plants. A sweet water body is located adjacent to the beach at the base of the hill. Mandrem beach on this sector also has long stretch of sandy shore with mature dunes and well-protected sand dune vegetation. Here one can clearly see the embryo Dunes, fore and Mid Dunes and Mature Backshore Dunes (Mascarenhas, 1998). The entire coastal stretch is protected by continuous 5-6m high sand dunes with dune vegetation. Some of the sand dunes are shifting in nature as can be observed in Mandrem where new dunes, presently low in height, are being formed thus making the coast prograding.

Ashvem (15.6^0 N / 73.7^0 E) slightly to the south has a mixed shore with sandy and rocky beach. The coastal dune vegetation is well protected in the area.. Morjim also preserves a good sandy shore with well-developed dune system. A long strip, which terminates at the mouth of river Chapora is marked by extensive rows of sand dunes with dune vegetation being more pronounced in the southern part. The back dunes are not well stabilized but the embryo and fore as well as mid dunes are relatively better preserved.

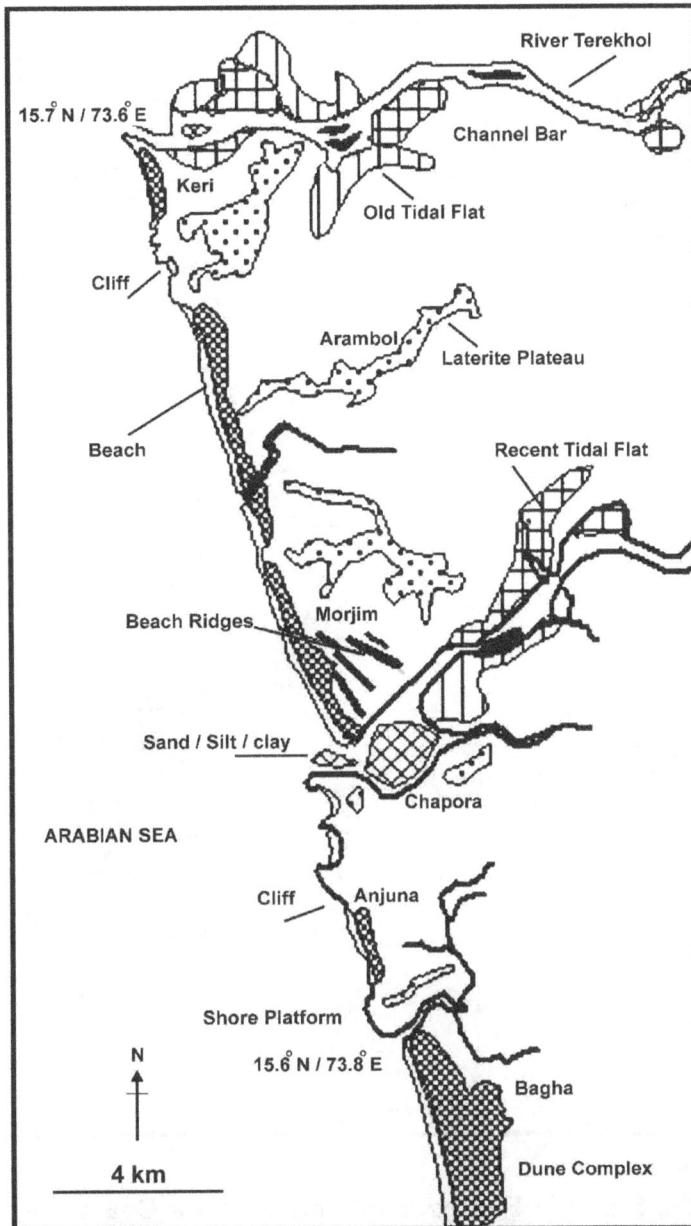

Fig. 44 : Features of North Goa Coast

South of Chapora fort, and especially at Vagator, low sand dunes are common over a limited stretch. It comprises of a rocky coast with some sandy pockets.Similarly, at Anjuna, there are extensive sand dunes covered by vegetation. Many dunes are found to be degraded and altered. One can clearly see many disturbed chaotic strandlines at several places along the shoreline in this sector (Mascarenhas, 1998).

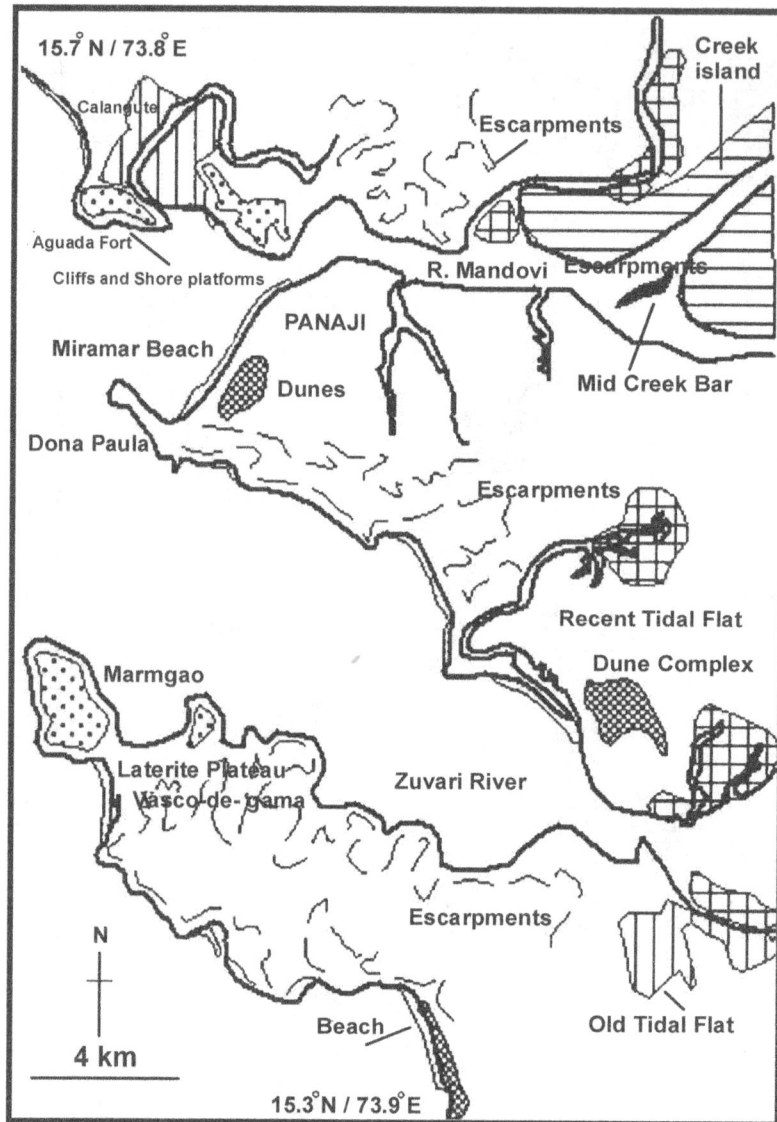

Fig. 45 : Features of Central Goa Coast

Central Coast : The Mandovi and Zuari estuaries in Central Goa form the largest tidal system along Goa coast (Fig.45). The large embayment at the entrance of the Mandovi estuary, Aguada bay has resulted in very complex flow patterns of the tidal flow. The main mechanism causing siltation in the channel is deposition of suspended sediments and sand movement due to littoral drift (Kunte *et al*, 2001). Mudflats are found mainly along such estu-aries and creeks. It is hypothesised that the promontory on which Vasco-da-Gama lies is responsible for the formation of two different geomorphic zones. (Anand et al 1987).

Table 21 : Beach and Dune Systems on Goa Coast

Name	Location	BEACH		DUNES		
		Length	Width	Length	Av. Widt	Max. HT.
	Lat/ Long	(m)	(m)	(m)	(m)	(m)
	(N/E)DD					
Keri	15.71/73.69	1900	41	800	262	8
Arambol (N)	15.69/7a69	418	49	113	18	6
Arambol (S)	15.67/73.7	3460	110	2021	235	6
Mandrem	15.66/73.71	3747	140	3500	194	5
Morjim	15.62/73.73	2502	82	1325	210	4
Vagator	15.60/73.73	760	72	300	105	7
Anjuna	15.57/73.74	1320	37	1300	120	7
Calangut	15.54/73.75	7300	66	1100	320	9
Miramar	15.47/73.80	3060	86	2000	169	3
Colva	15.24/73.92	26636	67	25000	460	5
Agonda	15.04/73.98	2870	69	2600	136	6
Palolem	15/74	1440	50	1100	87	7
Canacona	14.99/74.03	1960	94	900	340	6
Polem	14.9/74.08	670	46	290	117	6

Calangute beach is the longest beach on this section and has a length of about 7 km. It is backed by a one km long and 6 m high dune zone which stretches between Bagha in the north and Candolim in the south. This is a 300 m wide zone that comprises of foredunes, backdunes and interdunal areas. Miramar beach is another important beach which is 3 km long and 85 m wide. Dune zone at the back of this beach is very low. The headlands at Aguda fort, Dona Paula and Marmgao are 72m, 35 m and 75 m high and show sea cliffs having height of 10 to 12 m above sea level and narrow shore platforms with a width rarely exceeding 10 m.

The Baga – Calangute stretch of central Goa coast consists of sand dunes, with diverse dune vegetation. A large number of dunes are severely altered, leveled and eliminated due to anthropogenic activities. At Calangute, sand dunes, with dune vegetation and casuarinas, in various stages of degradation are found to its north and south. The Calangute Sinquerim stretch up to the Aguada headland is marked by a continuous chain of 5 – 6 m high sand dunes with diverse dunes and vegetation cover. Anjuna, Baga, Calangute and Candolim are most affected beaches due to tourism.

A prominent sand dune belt is found backing the Miramar beach. New dunes, presently low in height and capped by vegetation, are in the process of formation along the

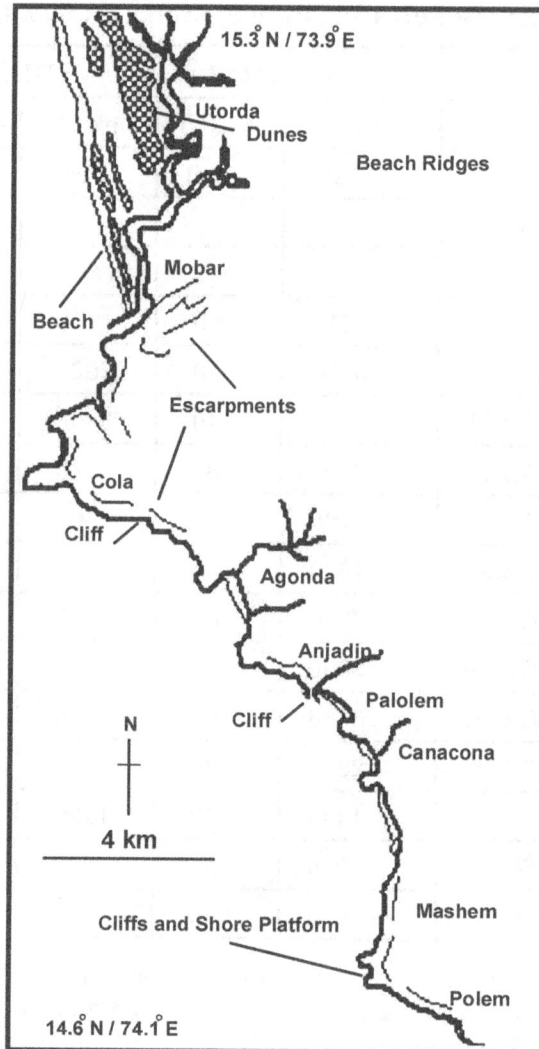

Fig. 46 : Features of South Goa Coast

northern part of Miramar beach; it terminates against the wooded hill slopes of Cabo promontory (Mascarenhas, 1998). However, sand dunes are progressively being destroyed. Due to severe erosion, the primitive beach no longer exists, except for a small portion near Panjim city. Instead, a sea wall, almost 2 km long, has virtually replaced the beach; this concrete / stone wall is placed in the intertidal zone. The Vasco to Sancoale (15.4^0 N / 73.9^0 E) sea front comprises of rocky headlands. It has only two major beaches: the one at Baina and the other at Bogmalo Here, low sand dunes are present but some of them are obliterated. To the south, the coast is mostly rocky (Chicolna) with a few pocket beaches and secluded cover such as the one at Hollant.

The Southern Coast : It is characterized by 18 to m high sea cliffs with very narrow shore platforms. Along the southernmost coastline shore platforms are literally absent.

Along Colva beach (Pale – Utorda – Mobar), 25 km long sand dune sector is fronted by 26 km long sandy beach (Fig. 46). Dune system comprises of 4 to 6 m high fore and back dunes.Dunes are also found along Agonda, Palolem Canacona and Polem beaches.Dune sector along South Goa coast varies in width from 115 m to 450 m.The beaches in this part of Goa are usually less than 3 km in length.

These beaches to the south indicate the prograding shoreline with deposition being the dominant phenomenon. To the south a wide regular beaches backed by a palaeo strandline suggest an earlier high sea in Holocene.

Sancoale to Mobor is a linear stretch with a very wide beach, backed by the largest and the longest strip of sand dunes of the entire coastal zone of Goa (Mascarenhas, 1998). At Velsao, sand dunes are very low. Most of the shoreline is under threat of erosional processes as evidenced by eroded berms and uprooted trees. The Cansaulim – Arossim stretch consists of very prominent 6 to 8 m high vegetated sand dunes. The Utorda – Majorda – Gonsua – Betalbatim coastal zone is marked by long strips of sand dunes, some as high as 8 m, with associated vegetation. At several places in this stretch, sand dunes have been flattened and obliterated. Severe beach erosion is observed near Majorda beach resort (Mascarenhas, 1998).

At Colva, sand dunes, 5 to 6 m in height, are seen. A small seasonal lagoon along the beach is an annual hydrodynamic feature at Colva. The Sernabatim – Benualim – Varca – Cavelossim – Mobor coast is backed by protective sand dunes, varying in elevation from 3 to 10 m. At Cavelosim and particularly Mobor sand dunes are severely damaged. The stretch from Velsao to Mobor is characterized by the most majestic sand dunes and is the most exquisite dune belt of the entire coastal zone of Goa(Mascarenhas, 1998).

The Talpona (14.9^0 N / 74^0 E) beach area comprises of prominent sand dunes. There are thick mangroves within the estuary to the north and a headland with a thick forest cover in the south. The entire belt is characterized by majestic sand dunes, some as high as 6 – 8 m.

Agonda, Palolem and Galgibag have a very good fore dunes. Loliem and Polem have small strips of sandy beach very close to nearby village.

Beaches on Goa Coast :

Many of the beaches on Goa coast are bounded by headlands. The rocky headlands are totally independent of the formative beach processes and have a major influence on beach plan form, sediment transport and the morphodynamics (Karlekar, 2015). They are transformed to narrow beaches with steep to very steep beach faces in monsoon with varying degree of steepness and beach cutting. The maximum morphological changes occur during early monsoons (June—August). During this period most of the material is transported to the offshore and some alongshore. These beaches accumulate maximum sediment storage during April/May. They are then subjected to rapid erosion with onset of SW monsoon winds and wave conditions, followed by slower rates during the subsequent period of the monsoon. Erosion continues till August, when the beaches have minimum sediment storage. The wave climate during post monsoon and winter months helps the beaches in recovering gradually after passing through a secondary phase of erosion associated closely with the onset of northeast monsoon during November/December (Kunte et al,2001).

There is a considerable amount of variability in sandy beaches, which is a result of wave environment. The entire beach zone consists of depositional facies formed by waves; wave induced currents and associated flows.

The length and spacing of Goa beaches is entirely dependent on pre existing topography. Along the hilly coast the average length of beaches is less than 3 km. Most of the beaches are bounded by headlands at both the ends. Some beaches have a partial headland located down drift of the main headland (Moreno et al, 1999). Moreover there are embayed beaches which represent relatively stable sections of the coastline and typically have a curved plan form. The Miramar Caranzalem strip mostly consists of several pocket beaches backed by wooded hill slopes, now in various stages of degradation (Dona Paula). Notable sandy beaches are found at Vainguinim, Odxel, Bambolim, and particularly at Siridao where dunes are low. (Mascarenhas, 1998).

A striking characteristic of these beaches is the close correspondence between the beach plan form and the refraction pattern associated with the prevailing waves. If the embayment is in equilibrium with the hydrodynamic conditions, the refraction pattern is such that no net long shore transport occurs due to obliquely incident waves. This is because all breaking waves arrive normal to the beach along the entire embayment. Weak long shore currents are however recorded in some of the embayments. Rip currents can also be observed at the extremities of the embayment. The study of these embayed headland beaches shows that they have an asymmetric plan form characterized by a strongly curved shadow zone, mildly curved centre of the embayment and a relatively straight downcoast end. The shoreline at the straight section of the embayment is usually parallel to the dominant wave crests.

The average distance between the headlands controlling the embayed beaches in the area is found to be 2.4 km in a two headland situation. The maximum bay indentation distance is 0.7 km and the average wave obliquity angle along Goa coast is 23 degrees. The beaches are small pocket beaches with an average length of 2.4 km (Karlekar, 2015).

Estuaries : Out of seven major rivers in the state, only the Mondovi and Zuari rivers have developed large estuaries. The estuarine mouth of the Zuari River is abutted by prominent plateau heights of Bambolim and Cabo headland in North and Cortalim, Dabolim and Marmgoa in the south. The Mandavi estuary mouth is abutted by the plateau heights of Aguada in the North and to the South by the reclaimed portion of Panaji city, Miramar beach and Cabo.

The lower parts of estuaries are tidal where sea water encroaches during high tide and submerges the low lying areas adjacent to banks forming tidal flats. These are extensive, horizontal, marshy or barren tracts of land that are alternately covered and uncovered by the rise and fall of the tide. The sediments on the tidal flats consists of silt-sand or silt-clay, with abundant organic matter. The old tidal flats cover wide stretches along Terekhol, Chapora, Mondovi and Zuari rivers.

The Mandovi and the Zuari estuaries are about 5 m deep. Their cross-sectional area decreases from mouth to head, and tides occur in the two estuaries up to a distance of about 50 km (Sundar et al,2005). The runoff in the two estuaries is highly seasonal, just as is the precipitation.

The flow in the estuarine channels is primarily tidal after withdrawal of the monsoon, and continues to be so until onset of the next monsoon. The amplitude of the tide remains virtually unchanged in the channels, and the tide propagates from mouth to head with an average speed of about 6 m/s (Shetye *et al* 1995).

Estuarine Islands : Thee islands in the estuaries are rocky exposures and extend peripherally, because of heavy siltation. The Divar Island (15.5^0 N /73.9^0 E) is a typical example of this type of alluvial island. There are few nearshore island like island of Sao Jacinto (15.4^0 N / 73.8^0 E) just inside the estuarine mouth.

Sea cliffs and Shore Platforms : The cliffs are located along Goa coast at Anjuna, Vagator, Marmugoa, Cape Rama, etc. Abandoned cliffs have been also identified upto 1 to 2 kilometer inland. These are no longer experiencing wave attack as a result of relative drop of sea level. The isolated sea stacks found occassionally show lithologic and structural relation with coastal rocks, a little inland along the coast. At some places, the sea stacks appear worn down and are exposed only at low tides.

The shore platforms or marine abrasion platforms are found at the base of the cliffs and headlands. Along the northern sector of the Goa coast the platforms are cut in quartzites and laterite, while along the southern coast, these are in metabasalts and gness and granites. Along some beaches of the northern sector of Goa, conglomeratic laterite is exposed forming such sore platforms.

Near-shore Islands : There are a few islands off the Goa coast such as, Grande (15.4^0 N / 73.8^0 E), and St. Jacinto. They show a lithological and structural similarity with mainland. These islands are the detached portions of coastal headlands and promontories and are now isolated from the coast due to rise in the sea level or due to local tectonic movement.

(D) Coast of Karnataka

The coast of Karnataka state lying between Longitude 74°5' E and 74°51' E longitudes and Latitude 14°53' N and 12°45' N is about 280 km long and consists of Coastal stretch of Uttara Kannada, Udupi and Dakshina Kannada districts (Fig. 47). The coast is bordered by Arabian Sea on the west and Western Ghats in the east. The continental shelf of Karnataka has an average width of 80 km. Within 17 m depth off the Karnataka coast seven islands are located in a 10-km north-south stretch. They are Kangigudda Island, Kurmagadagudda Island and Shimisgudda Island with a maximum elevation of 32-61 m along northern coast; Karkalgudda Island and Mandalgudda Island with an elevation of 20-41 m off mid Karnataka oast; and, Mogeragudda Island and Anjadeep Island with an elevation of 13-46 m along the southern coast.

The coast is exposed to the seasonally reversing monsoon winds, average rain fall per year being 4209 mm. Of the total rainfall, 80% is received during June to August. The temperature ranges from 21Deg C in December to 36Deg C in April. The tides in the study area are mixed semidiurnal, the range of which increases towards the north. (Kumar *et al.*, 2011). The tidal range along the study region is about 1.5 m and the submergence of the land associated with high tide period is less than 5-6 m. During the monsoon along the coast, significant wave height up to 6 m has been reported (Kumar *et al.*, 2006), and is normally less than 1.5 m during rest of the period. (Hegde et al, 2015).

The coastline is sandy, and rocky cliffs are seen mainly in the vicinity of headlands. The coast comprises a narrow belt of coastal sand dunes, marshes, and littoral terraces. The coast is drained by the Kali Nadi, Gangavali, Bedti, Tadri, Sharavati, and Netravati

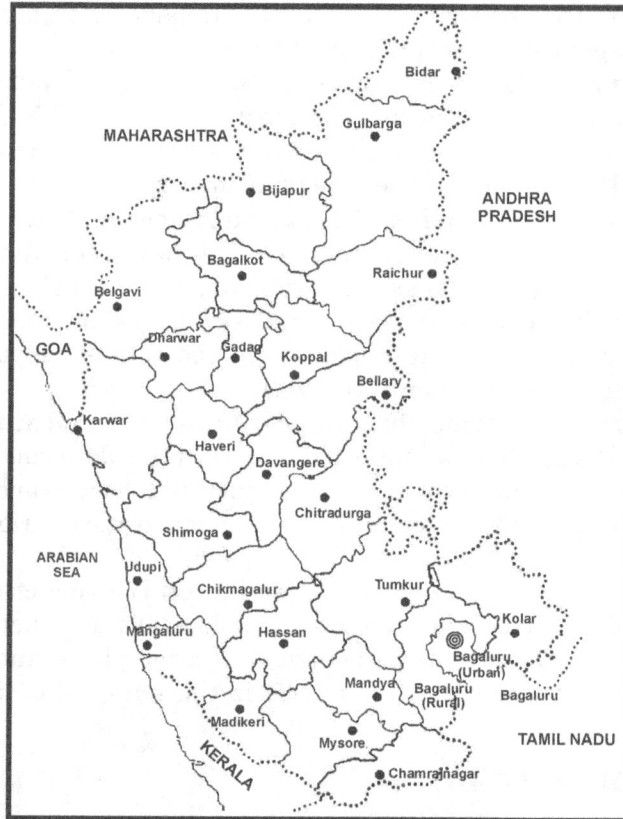

Fig. 47 : State of Karnataka

rivers, which have carved out narrow valleys with steep gradients and generally flow in a westerly direction. The major part of the Karnataka coast comprises of sandy beaches which vary in width from a few meters to 200 m.

The coast has preserved strong evidences of a submergent coast. (Nayak, 1993). Satellite image studies revealed northward shifting of the mouth of estuaries along this coast.

The coastline is intercepted with a number of rivers joining the Arabian Sea. Areas near the river mouths along the coastline suffer permanent erosion due to natural shifting and migration of the river mouths. Beach ero-sion is severe in some areas along this stretch. The erosion is not continuous all along the coast but is seen in isolated stretches. Relatively more erosion has been observed at the river mouths of Devbag (north of Kali River) and at Pavinakurve (north of Sharavathi River).

Highest accretion rate of 15.5 m per year has been noticed at Karwar whereas highest erosion rate was recorded at Honnavar with a value of about 19.59 m per year. It was also noted that the coast of Bhatkal is subjected more to erosion.

North Karnataka Coast :

It is charac-terized by pocket beaches flanked by rocky cliffs, estuaries, bays, and at

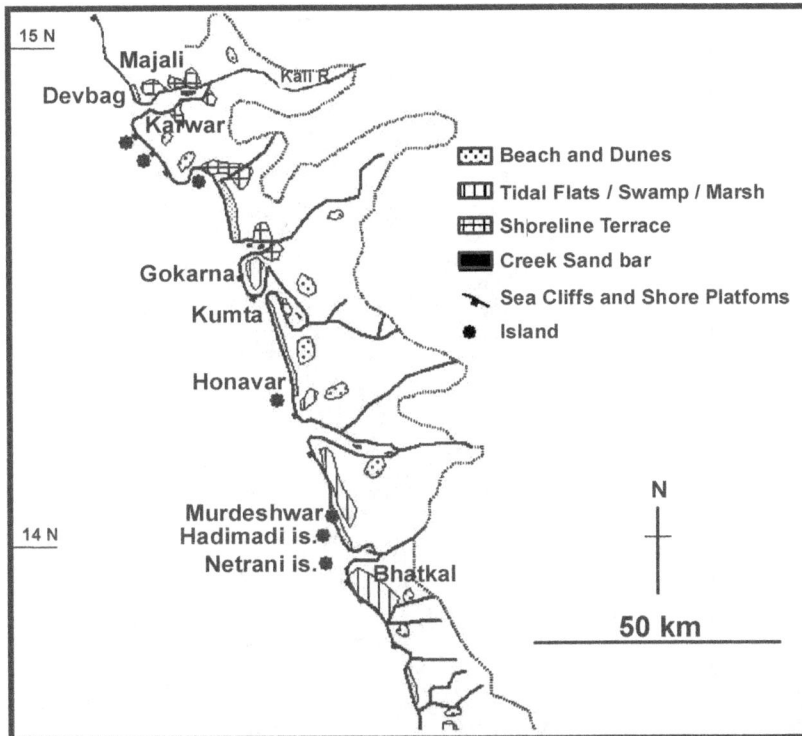

Fig. 48 : Features along North Karnataka Coast

some places mangroves (Fig. 48). Beaches of the North Karnataka occur in between estuaries and rocky headlands, Beaches in some places along northern Karnataka, however, are long and linear in nature with sand dunes. There are a few islands along the southern parts of this coast near Karwar. This coastal stretch is typical of a cliff coastline with raised shore platforms. Evidences of Palaeo strandline can be seen on this coast in the form of fossil beaches and dunes.(Fig.49). A number of sand dunes perched on rocky pediments and laterites, are also traced 4- 6 km inland from shoreline, around the Karwar town. Netrani on this coast is a 79 m high coral island which has a coral reef with many varieties of coral formations. The island is characterised by sharp rocks and steep cliffs. The coastal Quaternary deposits are flanked by the extensively developed multi-level laterite surfaces which descend in steps towards the Arabian Sea.

The northern coast of Karnataka is comparatively more irregular than the southern coast. Geomorphically they are therefore two distinct coastal stretches and comprise of different sets of coastal features in general. The circulation in the estuaries on North Karnataka coast is mainly controlled by river flow in monsoon and post monsoon months and by tides during the north-east monsoon and pre monsoon months (Kunte & Wagle 2001).

South Karnataka Coast :

It is mainly a straight coast characterized by long beaches and dunes (Fig. 50). The

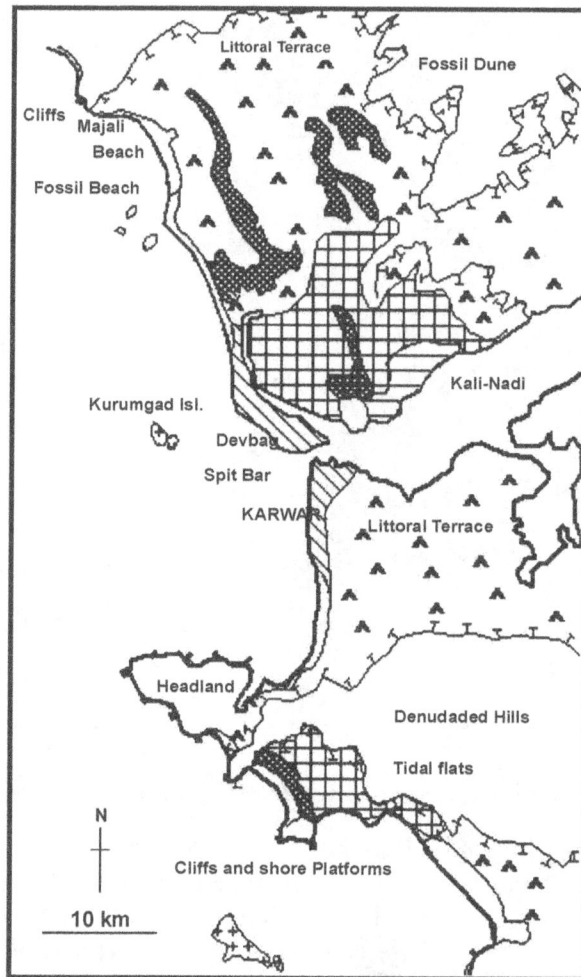

Fig. 49 : Features of North Karnataka Coast

beaches along this coast are almost straight, interrupted by a number of estuaries without headlands (Manjunath et al 1999). Kunte & Wagle(2001) have studied the southern part of coastline to understand the beach morphological changes. Their study showed that the beach erosion areas along the coast are migratory in nature. This migration is due mainly to the construction of breakwaters and seawalls on the coast. They concluded that there is a southerly littoral drift and any obstruction to this drift results in erosion. The net shore drift is towards south. Off Manglore, the alongshore transport of sediment in shallow regions is towards north whereas in the deeper regions, it is towards the south. This onshore-offshore sediment transport study revealed that the sediment transport direction is towards onshore where fluvial influence is absent and the direction is towards both onshore and offshore where the fluvial influence is prevalent Kunte & Wagle(2001). Studies on South Karnataka coast indicate that the beaches along this segment of coastline are essen-tially stable except some seasonal fluctuations.

The coastline at Malpe is almost stable with negligible erosion and deposition. Significant amount of loss of land is observed mainly at the river mouths due to the

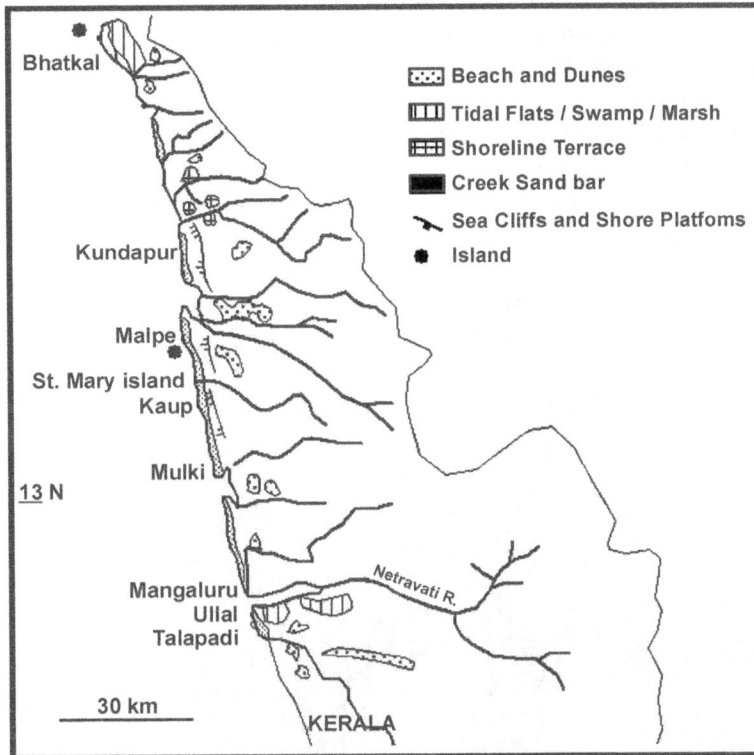

Fig. 50 : Features of South Karnataka Coast

sediment erosion from the banks because of complex interactions between river flow, waves and the tides. (Vinayaraj et al, 2011)

A significant spit bar complex has developed on the southern coast of Karnataka which consists of Mangalore spit to the north and Ullal spit to the south of Nethravathi-Gurupur river mouth. According to Raghavan et al (2001), morphologically, Mangalore spit has shrunk by 750 m in length between 1910 and 1993. The Ullal spit however shows an increase in length by 800.The spits show no prominent lateral migration. Shrinkage of the Mangalore spit and the growth of the Ullal spit indicates net northward migration of the estuarine mouth.

During monsoon season river discharge and consequent supply of the sediment to the shore drift is more. The Ullal spit consists of coarse to medium sand while the Mangalore spit consists of coarse to fine sand. As the Ullal spit grows towards north it induces the shift of river mouth towards north enforcing the erosion of Mangalore spit from its distal end. It was inferred that this is mainly due to the construction of a shore normal breakwater on the Mangalore spit (Fig. 51). Netravati and Gurpur rivers originate in the Western Ghats. These two rivers unite to form a common estuary before debouching into the Arabian Sea. According to Majunath et al (1999) observation of lithologic succession in a number of drilled boreholes and dug wells in the area indicate that in last glacial period these rivers had independent confluence points with the Arabian Sea. The submergence of river channels and growth of 8 km long barrier spit under the strong influence of southerly littoral currents are responsible for unification of Gurpur river with the Netravati before they drained into the Arabian Sea during the late Holocene.

St. Mary's Islands, also known as Coconut Island and Thonsepar, are a set of four small islands off the coast of Malpe in Udupi on south Karnataka coast. They are known for their distinctive geological formation of columnar basaltic lava. Out of the four islands, the northernmost island has a basaltic rock formation in a hexagonal form, the only one of its type in India. The island covers an area which is about 500 m in length with a width of 100 m. The north-south aligned islands form a non-continuous chain. The four largest islands are Coconut Island, North Island, Daryabahadurgarh Island and South Island.

The islands are generally aligned parallel to the coast line. The islands' terraces and elevated beach deposits are the proof of the reported fall in sea level of about 1 mm/per

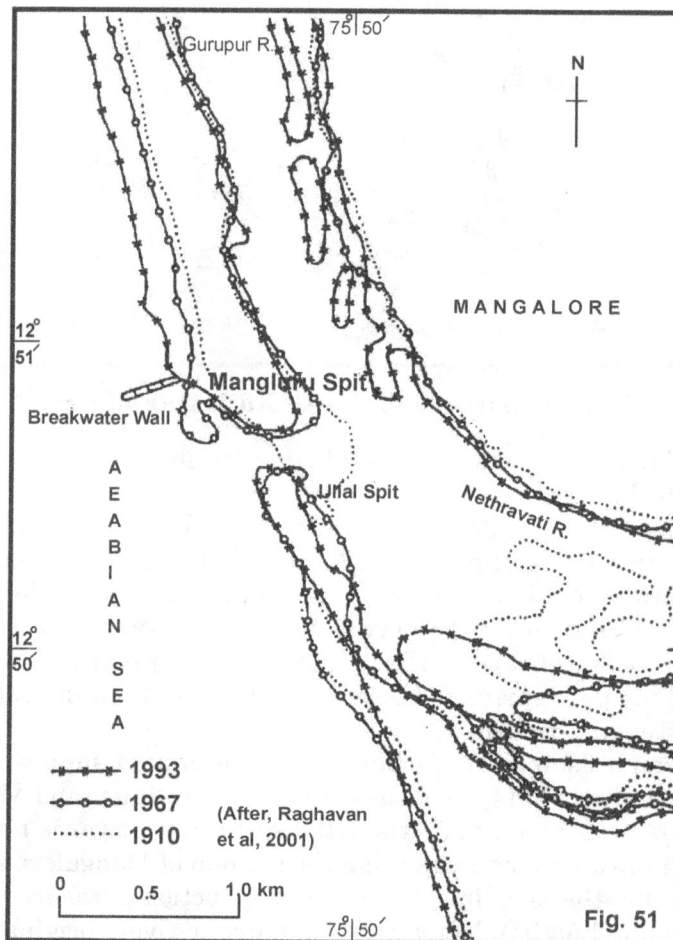

FIg. 51 : Morphological Change in Mangaluru-ullal Spit Complex

year. The highest elevation at Coconut Island is about 10 m above msl. It is surrounded by platforms in the elevation range of 1 to 6 m which also suggest higher sea level in Holocene.

In Karnataka, according to ministry of environment and forests (2001) mangroves are sparsely distributed in the estuarine areas. The substrate is made of fine-grained clay particles and is rich in nutrients. The condition of mangroves is relatively better along the Mulki and Sita-Swarna rivers; and in the Chakra-Haldi-Kolluru riverine complex, Sharavati estuarine complex, Tadri creek, Aganashani riverine complex and Kalinadi estuarine complex. Rhizophora, Avicennia, Sonneratia and Acanthus are the common varieties of mangroves found in Karnataka tidal inlets. There is a dense growth of mangroves in Kundapur region. Mangrove trees with the height ranging from 10-12 m are found in this region.

(D) Coast of Kerala

Kerala coast is a 560 km long narrow strip of land bordering the Arabian Sea. It extends from 8° 15' N to 12° 85' N latitudes and 74° 55'E to 77° 05' E longitudes and shows a remarkably straight coastline oriented in NNW - SSE direction. It is believed to have originated as a result of faulting during the late Pliocene (Krishnan, 1982).

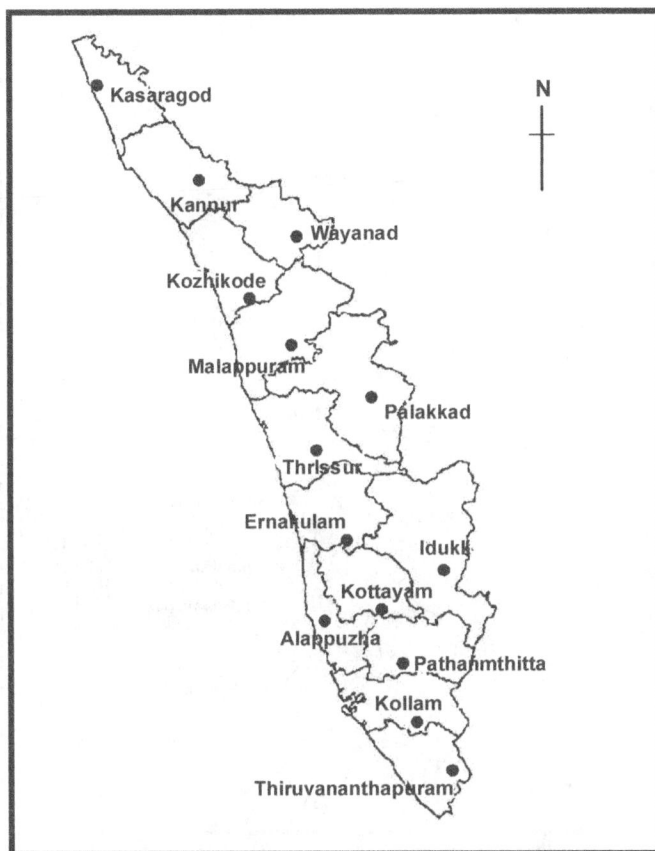

Fig. 52 : State of Kerala

Kerala plains are much wider and less hilly than the rest of the west coast. Recent observations indicate that the shoreline as a whole is dynamic and neotectonically active leading to considerable erosion and loss of surface area. Of the 560 km shoreline of Kerala, a cumulative 360 km length of shoreline is vulnerable and shows wide fluctuations in its stability (Nair, 1987). Narrow stretches of pocket sandy beaches are present all along the coastline. The major rivers that originate in western Ghats and flow west towards the coast are Pamba, Periyar, Bharathapuzha and Chaliyar. The coast is also known for well-developed bars, spits, low cliffs alternating with pocket beaches, barrier beaches, promontories, head-lands and bays.

The width of the continental shelf varies from place to place along the Kerala coast. It is wider on the northern side and narrower on the southern side. For example, the

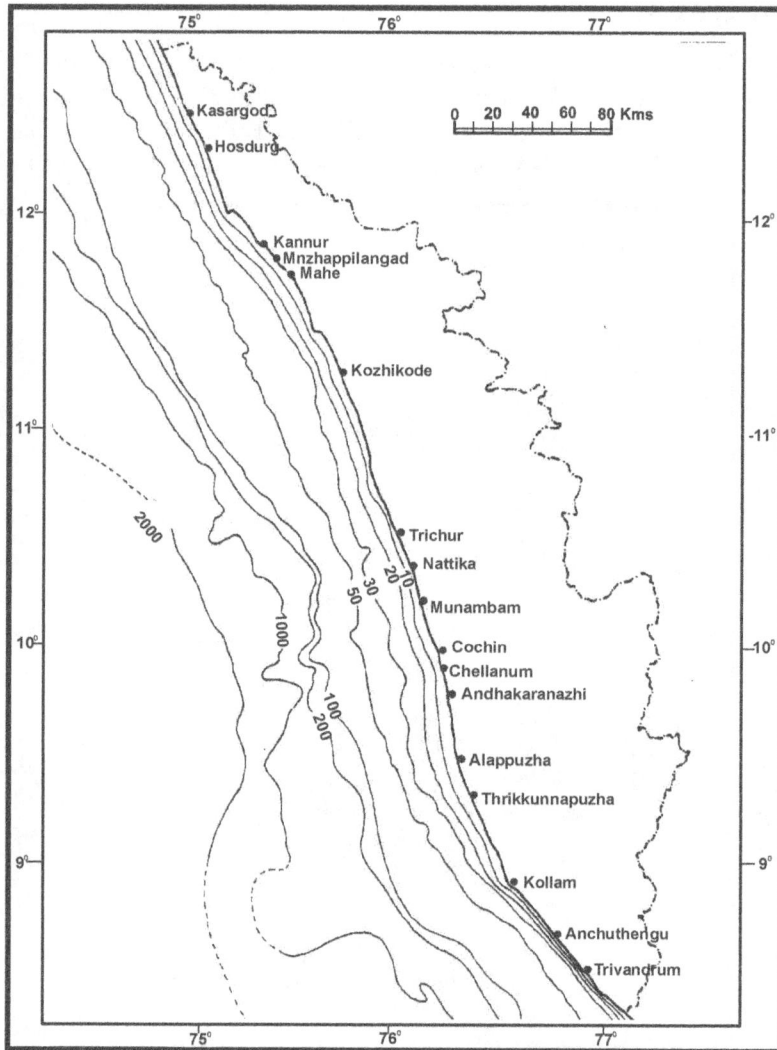

Fig. 53 : Bathymetry off Kerala Coast

width of the continental shelf at Kozhikode(11.3⁰ N/ 75.8⁰ E) is 80 km and it is only 45 km at Anchuthengu (8.9⁰ N / 76.4⁰ E). Along the southern side, the different depth contours are concentrated close to the shore. (Fig. 53) The slope of the continental shelf decreases towards north and increases north of Kannur (Baba, 1988). Muddy bottom shelf extends 50 to 60 kms from the coast to a depth of 100m. Beyond this, the shelf slopes down steeply to 1000m. The bathymetry of the inner continental shelf and nearshore of the Kerala coast show considerable variability along its length.

According to Gopinathan (1974) Kurian has identified four different zones of bottom deposits in the shelf region. These are: (i) Sandy deposit in the near shore region up to a depth of 3.5 m. (ii) Muddy deposit with small quantities of sand beyond 3.5 m depth and up to 18 m line, from Mangalore in the north to kollam in the south. Off Cochin, the belt of grey muddy deposit extends up to 35 m. off kollam, however, the quantity of mud in the deposit becomes small and sand predominates. (iii) Sandy zone which begins from the end of the muddy deposit and extends up to 100 or 120 m depth. In this zone, the quantity of mud progressively decreases and that of sand increases towards deeper water. (iv) Hard bottom zone begins from about 100 m line and extends up to 260 m depth. This zone has deposits of grey/black and white sand mixed with fine shell fragments and contains very small percentage of silt. Occasionally patches of rocks also occur in this zone.

The latest fishing charts of the coast show that the area between long. 76°40'E and long. 77°00'E and lat. 7°30'N, and further north of it along the shelf, is not only devoid of mud but it is hard and covered with sand and rocks. Probably strong currents near the bottom have prevented the natural deposition of mud in this region.

The Kerala coast is known for the presence of laterite cliffs, rocky promontories, offshore stacks, long beaches, dunes, estuaries, lagoons, spits, and bars.Three sets of sand dunes have been identified on this coast. Along the coast, sand ridges, extensive lagoons, and barrier islands are indicative of a dynamic coast. About 410 km of the 560 km coastline is protected by seawalls and about 30 km of the coast is undergoing severe erosion. Maximum loss of material has been reported along the southern sections.

The tides of Kerala are mixed, semidiurnal in nature and occur within the microtidal range of less than 2 m. During south-west monsoon, due to the strong winds, increase in wave activity with long swells and high breakers have been observed along the Kerala coast. Significant wave approach directions vary between 180° - 340° due to the NNE - SSW orientations of the coastline. In the south-west monsoon season the predominant direction of waves is between WSW and WNW.

The direction of littoral current on this coast is towards north for waves approaching from 220°-240°, whereas, it is southerly for waves approaching 280°-300° (Kunte and Wagale,2001) Considerable sediment transport in the offshore direction and the trapping of these sediments in the offshore bar were confirmed by Kunte and Wagale.

They found that along various locations, longshore littoral transport is variable depending upon wave climate and availability of sediments. It is towards north as well as south depending upon wave approach direction and configuration of the coast. It was also noted that at almost all locations, onshore-offshore littoral drift follows the same pattern. During April/May (pre monsoon), on-shore littoral transport accumulates maximum sediment storage. With onset of SW monsoon, winds and wave conditions (during June-September), offshore littoral transport allows erosion at rapid rate and off-shore transport continues till August when beaches have minimum sediments storage. Later, the wave climate during post-monsoon season encourages on-shore littoral transport

Mud Banks :

Occurrence of mud banks is a phenomenon peculiar to the Kerala coast of India. The regions known for mud bank occurrence include (Fig. 53) the southern coastal strip (Thrikkunnapuzha. - Alappuzha), the central strip (Chellanum-Munambam) and the northern strip (Kozhikode-Muzhappilangadi). The mud banks are reported to be decisively affecting the equilibrium conditions thereby causing shoreline instability of the coast. They trap the littoral material from either side thereby preventing it's downdrift, causing accretion within the mud banks and erosion on down drift sides. The mud banks are unique transient nearshore features appearing during monsoons at Kerala. They are unique phenome-non occurring at particular locations along the Kerala coast during the southwest monsoon season, which act as natural barriers to coastal ero-sion.

Mud banks are clearly demarcated areas of calm water adjoining the Kerala coast, during the SW monsoon in the Arabian Sea (Gopinath, 1974). Mud banks occur as small elevations of consolidated mud throughout the year. During the SW monsoon, because of wave action, the fine mud particles get churned up into a thick suspension. A semicircular periphery then develops around the suspended mud in which wave energy gets consistently absorbed (Fig.54). After the monsoon, the suspended mud settles and gets consolidated. Those mud banks which become active almost every year are persistent types of mud banks.

The mere existence of mud in an area is not enough to form mud banks. The mud of the right texture must get consolidated at the right depth where wave action could churn

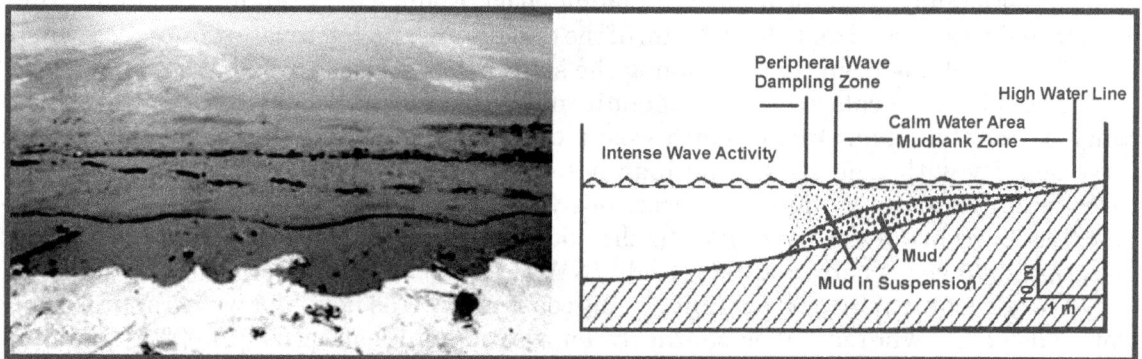

Fig. 54 : Mud banks

it up into a thick suspension. The presence of mud bank disturbs the shore stability of that region and induces coastal erosion in adjacent areas (Gopinath, 1974).

Mud banks, as they appear and disappear in the sea, have been considered as unique formations and seem to occur nowhere else except along the Kerala coast in India. The regions of 'the Kerala coast where these are noticed lie between Kannur and Kollam are generally known to appear prior to or during the SW monsoon, a week after the monsoon has set in, or immediately after the SW monsoon has commenced,.

The latest explanation for their occurrence is that their formation is related to the presence of rip currents in that area. The mud bank near Alleppey is one of the three persistent mud banks of the Kerala coast. The other two are found at Cochin and Kozhikode.

Backwaters :

The coastal zone of Kerala is well known for backwater systems, placer mineral deposits and mangroves. Between Kasargod and Kollam many long and irregular lagoons can be seen behind the impressive coastal barriers. Many of the lagoons, locally known as Kayals, are bestowed with numerous islands of different sizes. There are 34 Kayals in this area. Among these lagoons, Vembanad lake is the largest (205 sq.km) followed by Ashtamudi Kayal further south (Fig. 55). The Vembanad lake opens into the Arabian Sea at Cochin. Six rivers namely, Periyar, Pamba, Manimala, Achankovil, Meenachil and Muvattupuzha

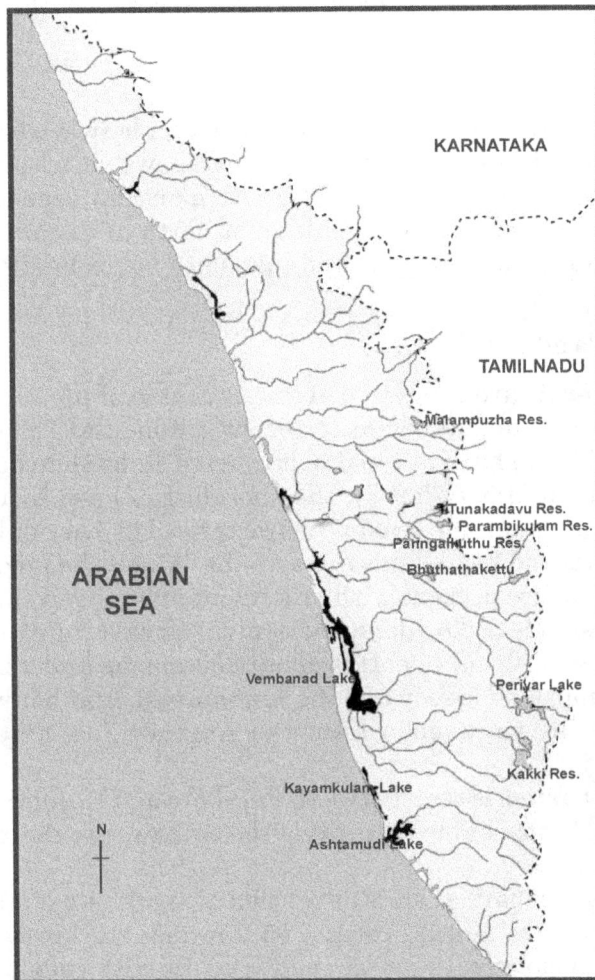

Fig. 55 : Backwaters of Kerala

discharge into this lake. Its depth varies from less than 1 m to 13 m. with a length of about 110 km and the width that varies from a few hundred meters to 4.5 km. Lagoons and estuaries play an important role in beach dynamics along this coast.They act as effective sediment traps along the coast.

Other group of backwaters include Kayals of Thiruvananthpuram-Kollam coastal stretch e.g. Vellayani Kayal, Veli Kayal, Kadinamkulam Kayal, Nadayara Kayal, Paravur Kayal, and Asthamudi Kayal. The evolution of the coastal lagoons has been influenced by the geological and geomorphological history of the coastal area and the sequence of changes in the levels of land and sea, which have resulted in coastal submergence and the formation of inlets and embayments. Subsidence of coastal regions may deepen and maintain coastal lagoons, delaying their infilling. The greater depth of the lagoon in the southern part, near Vaikom may indicate subsidence. The shallower depth in the northern part may be related to uplift. The morphology of the lagoon suggests that faulting has played a role in its evolution. The Vembanad lagoon at many places runs across the strandlines affecting their continuity.

Coastal Inlets :

They play an important role in the exchange of water between bays/lagoons and ocean. There are about 48 inlets in Kerala, out of which 20 shows permanent nature of opening, whereas the remaining 28 open only during the monsoon season (Nair et al., 1993). Munambam inlet is a major permanent inlet just north of Cochin inlet, through which Periyar River joins into the sea. Islands and inlets are major landforms that occur along the lagoon.

Beaches, Barrier Islands and Spits :

The beaches of Kerala are composed of fine to coarse grade sands (0.15 to 0.50 mm). The coastal area is mostly of sub-recent to recent sediments. The coast from Kollam to Koyilandy (11.4^0 N /75.7^0 E) have alluvial belt covered by laterite deposits.

North of Ponnani (10.8 N / 75.9 E), the shoreline is lined by continuous stretch of beaches. North of Trivandrum, the coast is characterised by barrier beaches except at few places where rocky cliffs and head lands are present. Where the lagoons open out into the sea across the bars, spits are present with or without submerged sand bars.

Several barrier islands occur along the central Kerala coast of which Vypin island (10.6^0 N / 76.2^0 E) is a significant one. It is about 25 km long and 2 km wide and acts as a barrier between the Vembanad lagoon and the Arabian Sea. Another wide, shore connected, barrier island separating the lagoon from the Arabian Sea, extends from Cochin to Cherthala.

A well developed barrier is seen between Pallikere and Kannur. Several barriers are observed near Kollam. Remarkable features of the barriers are their elongated formation with a very small width.

There are a number of barrier spits developed along the west coast of India. Of these, the longest one is along the Kerala coast (55 km long, 10 km wide) formed between Alleppey and Cochin, enclosing the Vembanad lake. Barrier spits are long, narrow sand accumulations that are generally parallel to the mainland and formed due to littoral transport of sand along a barrier spit connecting one headland to another and then migrating landward or seaward. The probable reasons for the formation of barrier spit along the Kerala coast is due to littoral transport of sediments particularly during the stable sea level condition.

Beach Ridges :

A number of parallel beach ridges (also called as strandlines) are observed at many places (Nair, 1987). The width of ridges varies from 50 to 150 m and the height varies from 0.5 to 200 m. These beach ridges represent successive still-stand positions of an advancing shoreline in relation to the sea. Occurrence of strandlines suggests that the coast has undergone a series of marine transgressions and regressions during the Late Quaternary. Significant changes were brought in the coastal configuration and associated landforms since mid Holocene period. A number of strandlines, running parallel to the coast for a distance of 15 to 25 km occur up to 15 km inland from the present shoreline and about 2-5 m above sea level in the central Kerala region. These strandlines are present on either side of the Vembanad lagoon.

The occurrence of these paleo-beach ridges suggests the progradation of coastal land. This may be either due to fall in sea level or rise in the level of land or both. It is possible that both fall in sea level and uplift of the coast have influenced the formation of cheniers/strandlines along central Kerala coast. The sea level was higher 3000 yr BP than the present day along this coast (Narayanan and Anirudhan, 2003).

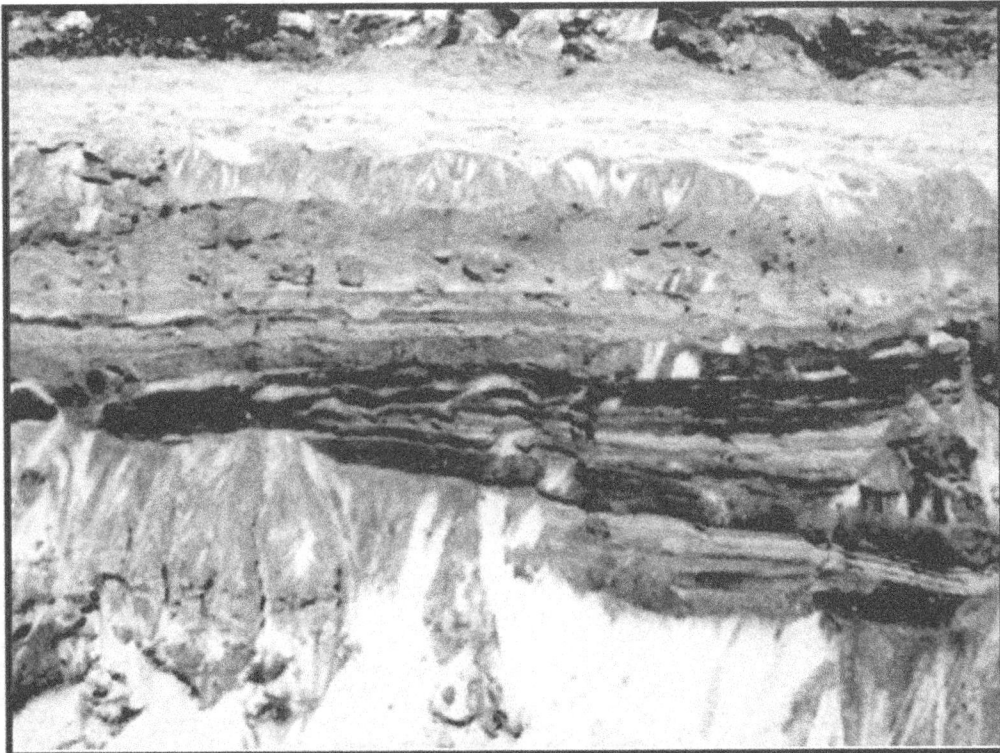

Fig. 56 : Placer Deposits in a Beach Cut

Mudflats, Tidal Flats and Mangrove Swamps :

Extensive tidal and mudflats are observed in the eastern part of Vembanad lagoon, particularly near the mouth of the southern branch of the Periyar river and in the Chithrapuzha river mouth area. Most of the mud/tidal flats are covered with mangrove vegetation.

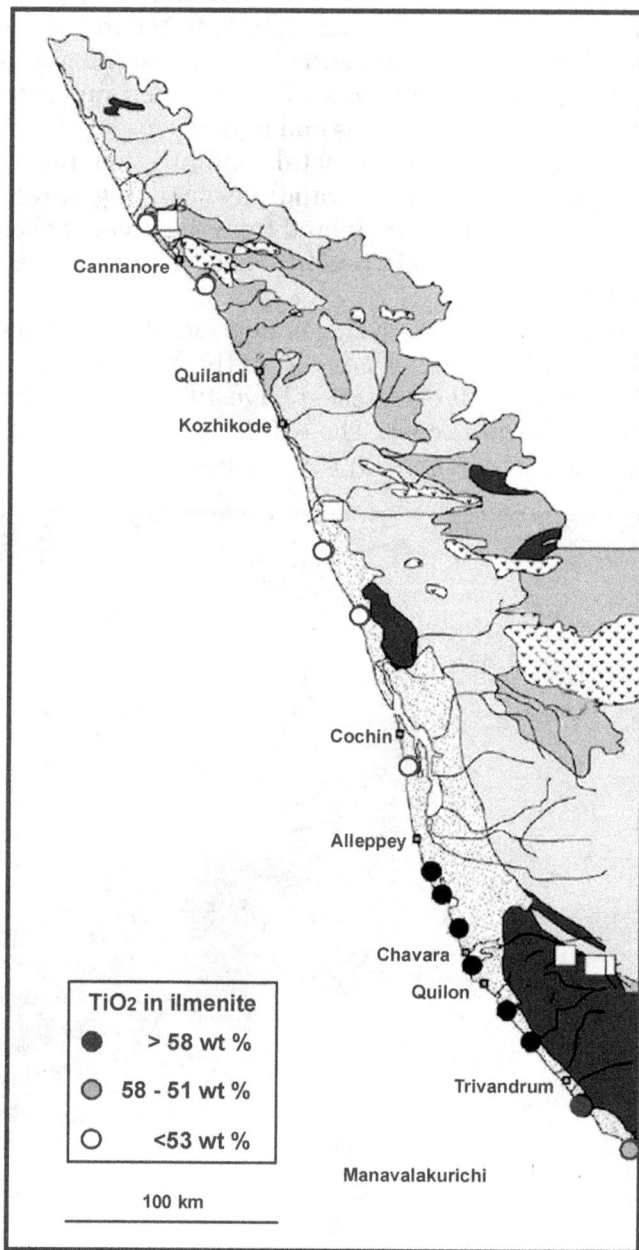

Fig. 57 : Beach Placers on Kerala Coast

Placer Deposits :

The placer deposits of considerable economic importance are present along the beaches of Kerala (Fig. 56). The concentration of the heavy minerals like Ilmenite, Monozite, Rutile and Zircon in the coastal area from Neendakara (8.9^0 N/ 76.5^0 E) to Kayamkulam (9.2^0 N / 76.4^0 E) is an important feature of the coast. Apart from various shades, the beach material comprises of shell fragments, magnetite, sillimanite and rare earths. Beach placer deposits are characterised by diverse nature, mineral assemblage, concentration and tonnage. The variations are attributable to the geologic, geomorphic, climate, tectonic, structural, biotic and hydrodynamic regimes.

The state of Kerala is by far the best in India, in terms of titanium mineral placer resources especially of ilmenite (Fig.57). Beach placers of Kerala coast contain 7-64% total Heavy Mineral in it. The geomorphic features of the south Kerala favour excellent formations of placers on beaches.The localisation of placers on this coast is mainly controlled by long-shore drifts in the area.

Sea Cliffs :

Along with long barriers and narrow beaches the Kerala coast is also characterised by steep cliffs. The cliff sections on the southern coast comprise both permeable and impermeable rocks, whereas those along northern coast are comprised of either Precambrian crystalline and/or Tertiary formations (Kumar et al, 2009). Cliff slope failure in the form of mass wasting, mudslide and mudflow are common in permeable rocks, Notches, caves and even small arches are developed in Kannur, Dharmadam and Kadalundi cliffs, where primary laterites are exposed to wave attack. Stacks composed of laterite and Precambrian crystallines found in nearshore of cliff coast indicate recession of shoreline. Rotational sliding, rockfall and toppling failure are found in hard rock cliffs.

Retreat of sea cliffs is induced by natural or anthropogenic activities or both. Rate of recession varies from a few centimetres to one metre/year depending upon the nature of lithology, structure and agents of erosion acting upon the cliffs. Narrow shore platforms are seen at few places like kannur.

Chapter 5
EAST COAST OF INDIA

Introduction :

The East coast of India comprises of coast of Tamilnadu, Andhra Pradesh, Odisha and west Bengal. Like west coast, coastal stretches of these states are also affected by specific coastal climate comprising of winds, waves and tidal and littoral currents as well as lithology and fluctuating sea levels.

Coastal Climate :

On the east coast, wave activity is significant both during southwest and northeast monsoons. Extreme wave conditions occur under severe tropical cyclones, which are frequent in Bay of Bengal during the northeast monsoon period. On the east coast, waves approach from southeast during southwest monsoon and fair weather period, and from northeast during northeast monsoon. Along the east coast, longshore transport is southerly from November to February, northerly from April to September and variable in March and October (V. Sanil kumar *et al*, 2006). The coast between Pondicherry and Point Calimere in Tamil Nadu coast experience negligible quantity of annual net transport. Seasonal variations in atmospheric pressure, relative humidity and wind speed along east coast is shown in table 22.

Tides :

Table 23 shows the average tidal range at important coastal sites. The coast experiences mixed semi diurnal tides with a tidal range that varies from less than 1 m to more than 6 m. The tidal range gradually increases from south to north i.e. from 0.8 m at Nagapattinam to 6.4m at Ramchandrapur. Tidal currents are very weak along the coast of Kannirajpuram (9.1^0 N /78.4^0 E). Here their velocity rarely exceeds 20 cm /s. Tidal currents with a velocity of 50 to 60 cm per second are recorded along the coast of Nagapattinam (10.8^0 N /79.8^0 E). The Southern coastal stretch up to Gopalpur enjoys micro tidal environment where tidal range is less than 2m. The stretch from Chandball to Malta river entrance is macrotidal with a tidal range varying between 2 to 4 m. Stations north of Malta river entrance and Paradip experience the tidal range of more than 4m. Tidal mean ranges are usually higher in semi-enclosed seas and funnel-shaped entrances of bays and estuaries and are typically low on the open coast.

Table 22 : Weather and Waves on East Coast

Parameter	Thoothukudi	Pamban	Ammapatinam	Salipetta	Sompeta
Pressure (MSL) mb					
April (pre monsoon)	1010	1011	1010	1011	1011
July (monsoon)	1008	1008	1008	1004	1000
November(Post monsoon)	1013	1012	1012	1014	1016
Relative Humidity (%)					
April (pre monsoon)	73	77	76	78	79
July (monsoon)	60	79	80	90	89
November(Post monsoon)	78	83	77	77	79
Wind Speed (knots)					
April (pre monsoon)	15	13	8	6	6
July (monsoon)	19	17	8	5	5
November(Post monsoon)	6	8	5	4	4
(Source: NHR, 1981)					

Sea level Scenario :

Along the east coast, sea level studies have mostly been restricted to de-positional and erosional features like inland beach ridges, rock terraces, cliffs and caves, shore platforms and submerged valleys. The available information is qualitative and comprises descriptions of strandline generated features. There is however no doubt that with the help of radiometric dates available now, it is possible to arrive at a reasonably good picture of sea level behavior along the East coast of India.

Banerjee and Sen (1987) investigated the sub-surface Holocene sediments around Calcutta from paleobiological point of view, and reported existence of mangrove forests about 6000 yrs.BP. Niyogi(1971) reported three terrace levels 6.1, 4.7 and 3.8m above MSL in the Subarnrekha river delta and the adjoining coast of West Bengal, and according to him, the entire delta with all its features was post-Pleistocene. On the Odisha and Andhra coasts which are better investigated, a higher strandline extending 20 to 30km inland and of the order of + 10m high is postulated (Prudhvi Raju et al 1978, Vaidyanadhan 1991, Merh,1992).

Table 23 : Mean Tidal Range on East Coast

Station	Lat (DD)	Long (DD)	TR (m)
Thoothukudi	8.78	78.15	1
Pamban	9.27	79.20	1
Nagapattinam	10.75	79.83	0.8
Cuddalore	11.73	79.78	1.3
Chennai	13.08	80.28	1.5
Pallamkuru	16.58	82.30	1.7
Kakinada	16.93	82.25	1.7
Vishakhapattanam	17.67	83.27	1.8
Gopalpur	19.23	84.92	1.9
Paradip	20.32	86.00	4.6
Chandball	20.73	86.83	3.1
Malta river ent	20.97	88.57	3.6
Sagar isl	21.63	88.03	5.8
Haldia	22.07	88.07	4.6
Ramchandrapur	22.17	88.17	6.4
Kolkata	22.53	88.32	6.2

Studies off Visakhapatnam coast of Andhra Pradesh (Prudhvi Raju *et al* 1978) have provided some data on high and low strandlines; features related to these sea levels have been chronologically arranged to indicate an early strandline + 7 to + 10m which regressed to —25m and then again arose to the present level (Prudhvi Raju and Vaidyanadhann, 1978). Perhaps — 25m strandline does not indicate the ultimate depth of the regression because the late Pleistocene low sea level all over the world has been of the order of —120 to — 150m. Other workers have report-ed submerged shorelines at various depths. Terraces and karst like features at —70 and -130m and carbonate reefs at — 50 to -60m provide evidences of sea stands at various stages of the Holocene transgression. These workers have suggested that the last transgression was post-glacial (Flandrian). The higher sea-stand according to them is related to an earlier transgression. Recently, Bruckner has stated that the Holocene sea along the Visakhapatnam coast was highest around 5100 ± 70 *BP.* and ac-cording to him the higher sea level reported by other workers was the last in-terglacial sea. A low sea-stand ⁻10 to — ll m has been provisionally dated 8000-6000 B.P. by Prudhvi Raju(1985)on the basis of drowned valleys and subma-rine topography.

Southward, around Godavari delta, Sambasiva Rao and Vaidyanadhan(1979) have recorded farthest strandline features (almost 35km inland) to be +8m while the nearest, +2m high (about 2km away from the coastline), and grouped them in Holocene. Subsequently

Bruckner has dated the levels within 18km from the present coast to be around 3600 ± 700 B.P. Comparable beach ridges (4 to 6m high) from Krishna river delta have been reported by others also. 14C dating of mollusc shell from a bar located at about 25 km from the present coastline in the Nizampatnam Bay adjoining western part of Krishna delta, gave an age of 8200 ± 120yr BP. suggesting a Holocene strandline of —17m. There ap-pears to have been a rapid rise in the sea level here around 8000 yr BP., when the barrier island was drowned and the shore zone migrated landward. Evidences of low sea stand at —49 and — 56m near Pulicat lake in the form of pebble horizons have been identified by Nageswara Rao 1978. A sequence of ridges from Cauvery delta, indicative of at least three strandlines, each rising 7.2, 6.9 and 5.5 above the present sea level have been described by Sambasiva Rao(1982). Earlier, Meijerink (1971) had invoked a low stran-dline —70m at the close of the Wurm glacial stage, which rose rapidly to the present level in the course of post-glacial transgression inundating 11000 to -12000 years old Pleistocene fluvial deposits.

According to Bruckner(1988), during the late Quaternary (Upper Pleistocene) transgression along Tamilnadu coast, the sea level arose between + 2 to + 8m during last interglacial stage and the Holocene transgression according to him reached its maximum during 6240 to 2740 years BP., the level hardly rising to 1m above the present level. The terraces around Mandapam and Rameswaram coast ranging in height above MSL from 0.62 to 0.20 meters, give ages varying from 5440 ± 60 to 140 ± 45.

(A) The Coast of Tamil Nadu

Tamil Nadu is situated in the southernmost part of the Indian peninsula with the Bay of Bengal to the east, the Arabian Sea to the west, and the Indian Ocean in the south. It has a long coastline of nearly 1100 km (Fig. 58) between Pazhaverkadu (Pulicat) (13. 4^0 N / 80.3^0 E) to Neerody (8.3^0 N / 77.1^0 E). A broad stretch of plains occurs parallel to the coast. The coastline is remarkably straight and narrow except for indentations at Vedaranyam (10.4 ^{0} N / 79.8^0 E) with well developed beaches, the most famous being Marina (second largest in the world). The Palk Strait separates Tamil Nadu from Sri

Lanka and is characterized by shallow water with sea grass beds.

The mean tidal range on the coast varies between 0.6 m - 1.5 m. and average wave height varies from 0.3m-1.8m. Tamil Nadu experiences a two-monsoon system viz. Southwest summer monsoon (June to September) and the Northeast monsoon (October to December). Coastal Tamil Nadu receives about 60% of its annual rainfall and interior Tamil Nadu receives about 40-50% of annual rainfall during the northeast monsoon. The Cauvery and its distributaries, the Palar and Tamarabarani are considered the major rivers of Tamil Nadu. The coastal area of Tamil Nadu is considered a rain-shadow area because of the low rainfall it receives during the Southwest monsoon.

At a few places mangrove systems, and at Gulf of Mannar and Rameshwaram fringing and patchy reefs, are seen. Deposition and erosion have been reported at different beaches along this stretch. Rich heavy-mineral deposits have been reported between Muttam (8.1^0 N / 77.3^0 E) and Manavalakuruchi (8.2^0 N / 77.3^0 E). The major landform along this coast is the presence of a large delta formed due to the Kauvery River and its tributary system.There are few more deltas also.

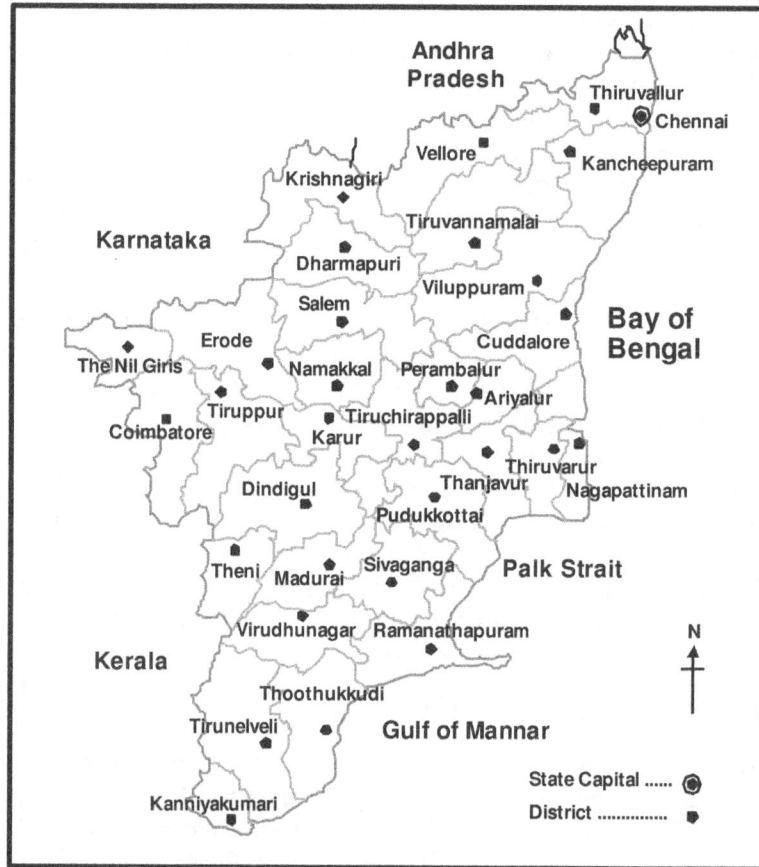

Fig. 58 : State of Tamil Nadu

The Deltas :

The Coast of Tamil Nadu shows a spectacular network of deltas between Chennai and Madurai formed by easterly flowing river systems (Fig. 59). It exhibits a typical fan shaped Proto Kauvery delta in the area west, northwest and north of Chennai with its apex located at about 45 km west of Chennai. It spans over an area of about 700 sq km. The axis of this delta is regionally oriented in NE—SW direction and the delta also exhibits a network of crescent shaped formations arranged as concentric rings (Ramasamy,1991) These are the depressions between two successive lobes in a delta front. It is a typical bird-foot shaped delta.

The Palar river near its confluence with the sea shows a distributory network of palaeo-channels. Throughout the year the river mouth is blocked by a bay mouth bar. Very few tidal creeks can penetrate the region.

Ponnaiyar delta close to Villupuram is a linear delta that occupies a narrow depression in uplands. Except the main channel all its distributories are abandoned and stand exposed as buried channels (Ramasamy,1991). In this delta there are narrow finger like sand bodies showing parallelism to the main channel suggesting that it is a tidal dominated delta

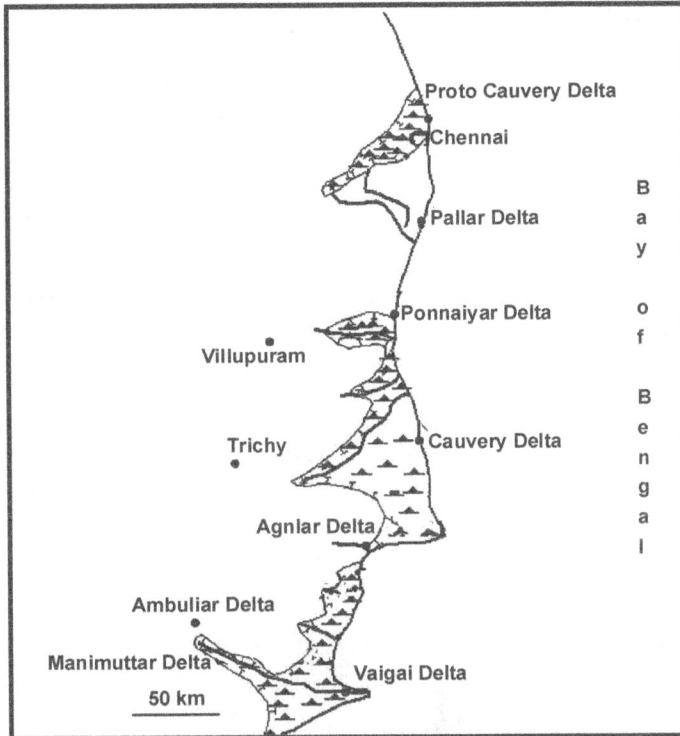

Fig. 59 : Deltas on Tamil Nadu Coast

The Kauvery delta is a triangular shaped delta with its apex located north of Thanjavur(10.8 °N / 79.1° E). All the distributaries show signs of abandoned river courses. The deltaic plain, the distributary channels and a flood plain are the main components of this delta.

According to Ramasamy (1991) marine interference and the strandlines are observed only to a width of 5 to 7 km from the present day shoreline.

The Agniar,Ambuliar,Vellar delta system is observed just south of the Cauvery delta The absence of continental delta and the presence of protruding cuspate delta is very striking feature.

The Vellar delta like Kauvery delta exhibits crescent shaped lakes arranged in a concentric fashion. It shows well developed paleo channels to the south. It also shows a protruding delta with a typical bird-foot shape

Vaigai delta to the south shows occurrence of beach ridges to a width of 2 to 4 km all along the coast (Loveson and Victor Rajamanickam, 1987). It shows signs of cuspate delta near its confluence. All these deltas were fluvially dominated in the initial phases of their growth but turned into wave dominated deltas later with tidal domination.

Mangrove Wetlands :

Table 24 gives the location of mangrove wetlands of Tamil Nadu and are shown in Fig 60. The major mangrove wetlands are located in the deltaic regions of the river Kauvery.

A large patch of healthy mangroves is present in the Devipattinam area, bordered by Palk Strait in the east, in Ramanathapuram District. In the islands of the Gulf of Mannar Biosphere Reserve, mangroves are present in a few hundred hectares. These mangrove patches consist of a true mangrove species namely, Phemphis acidula, which is not present in any other Indian mangrove wetland (Selvam *et al*, 2002).

Table 24 : Mangrove Wetlands on Tamil Nadu Coast

Location-District and estuary	Name of the Mangrove wetland	Area (ha)
Cuddalore: Uppanar-Coleroon estuarine region	Pichavaram	1357
Thanjavur: Coleroon estuarine region	Pudhupattinam	800
Thiruvarur-Thanjavur: Distributaries of Vennar	Muthupet	12000
Ramanathapuram: Islands of the Gulf of Mannar	Gulf of Mannar Marine National Park	30
Ramanathapuram: At the mouth of small tidal creeks at 11 places in the mainland	Palk Strait	700
Tuticorin: Tamirabarani estuary National Park	Gulf of Mannar Marine	148

Source : (Selvam et al, 2002)

The major mangrove wetlands of Tamil Nadu, namely Pichavaram and Muthupetare are located in Kauvery delta. They get their freshwater supply from the river Kauvery. The Pichavaram wetland is a vast plain with a very gentle slope towards the Bay of Bengal. The major coast features in the wetland are the beach, barrier dunes, estuary, tidal and mud flats, mangroves, spit/ tidal bar, beach terrace and strand line. The beach is very narrow with an average width of 50 m. The paleo shorelines have been found up to a distance of 15 km from the coast. A compound spit is seen at the southern tip of the Coleroon river mouth. It also helps in the development of cuspate foreland, which is triangular in shape (Selvam *et al*, 2002).

The Muthupet mangrove wetland is located near the southernmost tip of the Cauvery delta It is part of a large coastal wetland complex called the Great Vedaranyam Swamp. As in the case of the Pichavaram mangrove wetland, the quantity and duration of the freshwater inflow into the Muthupet mangrove wetland has reduced over the years due to the construction of dams and barriers in the upstream area, resulting in increased annual average salinity of both water and soil.

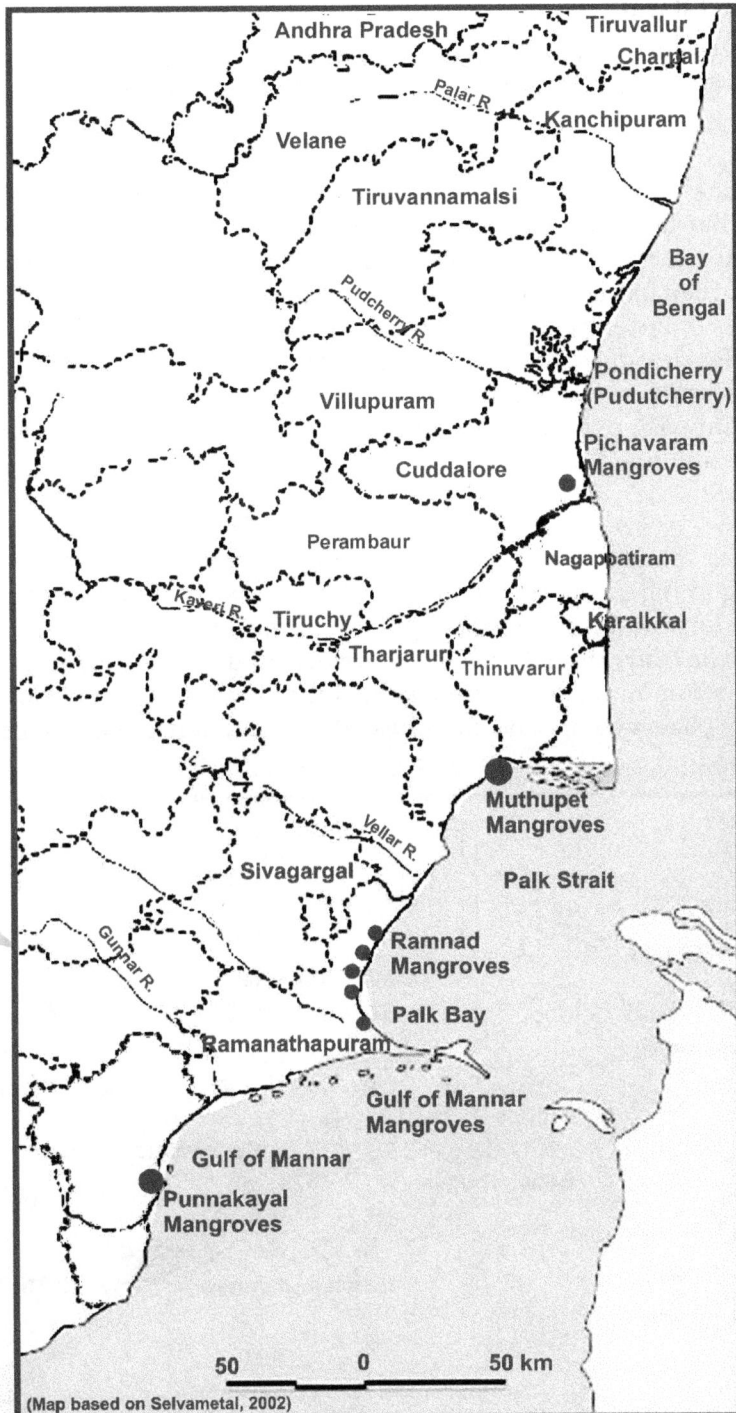

Fig. 60 : Sites of Mangrove Wetlands of Tamil Nadu Coast

Rameshwaram Island :

It is located on the eastern part of Ramanathapuram district of TamilNadu. It is bound between the latitudes E 79°12' 30" and E 79° 27' 30" and longitudes N 9° 8' 55" and N 9° 19'. It has an average elevation of 10 m. The island is spread across an area of 61.8 sq. Km·

Sea level has played a major role in the evolution of landforms on the island. The raised terrace that has been dated to 125000 years BP indicate that the there was only fringing reefs during that period (Prabkaran et al, 2010). According to them the emerged coral coast since then has been subjected to marine sedimentation as indicated by mudflats and beach ridges of latter period. The spit is the recently formed landform of the island.

Beaches, Beach ridges, Dunes, Lagoons, Mudflats, Creeks and spits are the major coastal features found on the island (Fig. 61).

The coastline also exhibits beaches, spits, coastal dunes, rock outcrops, mudflats, salt pans and strand line features

Beaches :

Beaches are found mainly in the southern and middle parts of the Tamil Nadu coast. According to Sathasivam et al (2015) the beaches on northern stretch of TN coast between Kalpakkam and Kodiakarai comprise of fine sand due to high energy condition existing there. The central coast beaches between Kodiakarai and Mandapam including Rameshwaram is dominated by medium sand due to low energy conditions and the similar pattern is also observed in the beaches along southern part from Mandapam to

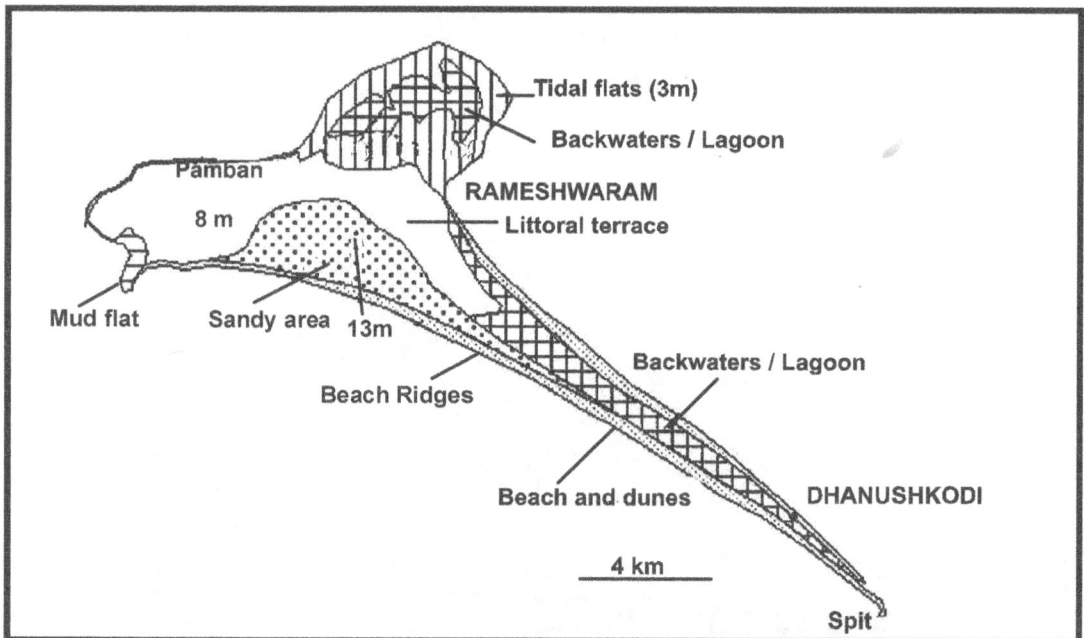

Fig. 61 : Coastal Features of Rameshwaram Island

Kanyakumari. The beaches and beach ridges around Tirunelveli, Kanyakumari and Manavalakurichi are rich in placer deposits (Loveson, 1993; Bruckner, 1988). The beach and dune sediments on the coast date back to Late Pleistocene to Recent (Chauhan, 1989). The beach sediments are well to very well sorted and characterized throughout by fine to medium grain size sands. A chenier is a continuous ridge of beach material built upon swampy deposits, often supporting trees.. These features are typically found in the northern central part of the study area.

Beach Ridges :

A beach ridge is a wave-deposited ridge running parallel to a shoreline. It is commonly composed of sand as well as sediment worked from underlying beach material. Different types of beach ridges are seen on this coast. They are usually associated with sand dunes. The height of a beach ridge is affected by wave height and energy. These features are found mainly in the northern and southern parts of the coast. Beach ridges are not seen in the middle part of the study area. Prominent beach ridge complexes are reported along the east coast, which are mostly occupied by large deltas; low lying hills and sandy shores. By contrast, West coast possesses less significant beach ridges (Kunte et al, 2004).

The beach ridges are generally formed at the limit of swash run up (Hesp 2000). On the Tamil Nadu coast between kallar and Vembar field ridges are barely discernable, typically being only 0.5 to 1.0 m high above the gentle undulating dune fields or low lying swampy plains lying sea wards (Chandrasekar *et al*, 2014). Ridges are typically parallel to the coast. Ridge and runnel are best expressed where the beach is completely devoid of coastal structures or terraces. The ridges consist of coarser sediments than runnel. The ridge and runnel topography normally shows slight longshore variability.

When the strandlines/beach ridges observed together, 3 distinct semi-continuous series of the beach ridges representing three strandlines can be noted on this coast(Kunte et al, 2004)

Sand Dunes :

Number of sand dunes of various dimensions are, developed along the fringe of this coast. The sand inputs between kallar and Vembar are rapidly recycled to the foredune where developing embryo dunes are consistently observed.

Sand Dunes are typically found along the southern central part of the coast between karaikal and kodikkari.Their average height hardky exceeds 3 m.

The estuaries and creeks

The estuaries are mostly occupied by rich mangrove forests, such as Pitchavaram (11.4° N / 79.8° E) and Pudukkottagam (10.4° N / 79.5° E).They can be seen mainly along northern coast from Chennai to Chidambaram (11.4° N / 79.7° E).

Mudflats :

Extensive mud flats are found along Vedaranyam Muthupet stretch of the coast. They are found in sheltered areas such as bays, bayous, lagoons, and estuaries on this coast as elsewhere on east coast of India. Geologically they are viewed as exposed layers of bay mud, resulting from deposition of estuarine silts, clays, and marine animal detritus. Most of the sediment within these a mudflats is within the intertidal zone, and thus the flats are submerged and exposed approximately twice daily. These features are found in the central to northern part of the coast. The PaleoTidal Flats which are ancient level muddy surfaces bordering estuaries are also found in the northern and northern central parts of the coast.

Sea Cliff :

Sea cliffs on this coast are limited in occurrence are not found frequently as in case of central west coast of India. They are restricted to small headlands and promontories on the coast.

A few cliffs with a height not exceeding 7 m are found mainly in the southern part of the coast near Kanyakumari. Cliffs and shore platforms are not the conspicuous coastal features on this coast.

The Coast of Puducherry :

Puducherry is a union territory situated on the Coromandel Coast between 11°45' and 12°03' N latitudes and 79°37' and 79°53' E longitudes. As far as coastal Geomorphology is concerned it is a part of Tamil Nadu coast and 24 km long coastal sector stretches between 11 degrees 47 minutes north latitude and 11 degrees 57 minutes north latitude (Fig. 62). The coast is endowed with a very wide range of coastal features such as estuaries, lagoons, mangroves, backwaters, beaches and dunes.

The Puducherry region receives its precipitation mainly during the northeast monsoon The tidal range on the coast, is low and the maximum range during a spring tide is around 0.8m. Recorded tide levels at Puducherry with respect to chart datum are: Mean High Water Spring (MHWS):+ 1.30 m Mean High Water Neap (MHWN): + 1.00 m Mean Low Water Neap (MLWN): + 0.70 m Mean Low Water Spring (MLWS): + 0.49 m (Ramesh et al, 2011).

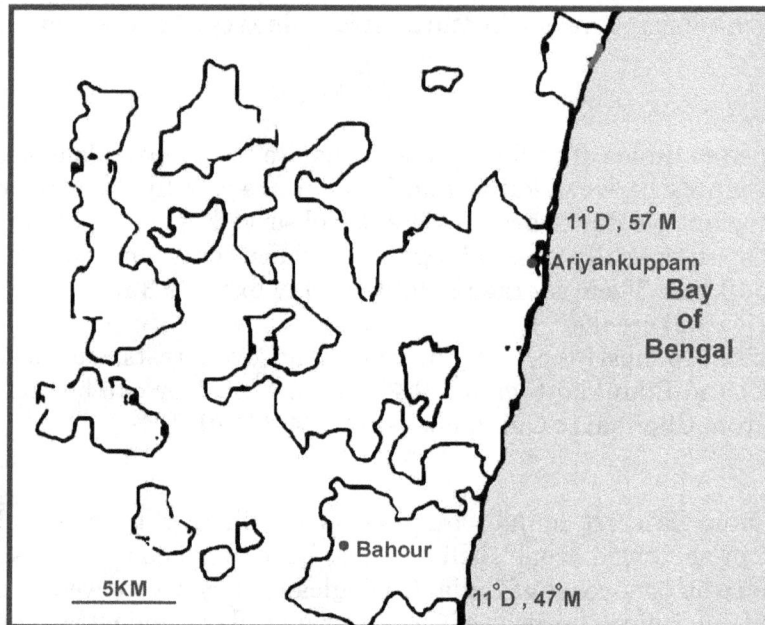

Fig. 62 : Puduchery

The tidal currents observed in the vicinity of Ariyankuppam River mouth indicate unidirectional current from north to south during flooding as well as ebbing tides and the maximum strength of the current as observed is 0.26 m per sec. At Puducherry, the wave heights rarely exceed 0.9m.

The terrain of this coastal sector is gently undulating with hight varying from 30 to 45m above Mean Sea Level towards interior northwest and northeastern parts of the region. The three major physiographic units generally observed are coastal plain (younger and older), alluvial plain and uplands.

Puducherry coastal region is a flat plain with an average elevation of about 15m above MSL. The coastal plain extends on the eastern part of the Puducherry region parallel to the Bay of Bengal. The sand dunes are seen extending all along the coast. Dunes and Barrier dunes are seen as continuous mounds between Ariyankuppam, and Manapet. Other features on the coastal plains are mudflats, creeks and tidal flats.

Along the Puducherry coast, beaches are generally narrow and severe erosion is observed along the northern segment of the coast. In the southern segment, beaches are comparatively broad and depositional (Ramesh et al, 2011).

(B) The Coast of Andhra Pradesh

The coast of Andhra Pradesh stretches from Pazhaverkadu (Pulicat) in south (13.5^0 N / 80.3^0 E) to Pata Sonapur in north (19.1^0 N / 84.8^0 E) on east coast of India.it has a total length of about 974 Km (Fig. 63).

The AP coast is under the influence of a microtidal environment. The coast near Krishnapatnam (14^0 N / 80^0 E) has a tidal range of 0.7 m, while near Vadarevu (15.8^0 N / 80.4^0 E) the tidal range is about 1.1 m. The coastline between Vadarevu and Machilipatnam shows tidal range from 1.1 m to 1 m respectively. At Vishakhapatnam to the north tidal range is 1.8 m. The normal spring high tide reaches up to about 35 km upstream in the distributaries of Krishna and Godavari rivers.

The coastline of Andhra Pradesh, mainly the deltaic coast from Kakinada to Vijayawada, is 300 km long and comprises of bays, creeks, extensive tidal mudflats, spits, bars, mangrove swamps, marshes, ridges, and coastal alluvial plains. To the extreme south, in the lagoon zone of Pulicate lake coastal inundations are seen. The deltaic coast has a dense cover of mangroves.

In the north, residual hills and ridges are seen close to the sea (Srirampuram,90 m, Rajapuram, 65 m, haripuram,92 m etc.) Small inlets and tidal mouths of streams are also seen here. Shore platforms, sea caves in rocks, and cliffs are seen at few places but they are not conspicuous coastal features as seen on west coast of India. Along the northernmost sector of coast in Ichchapuram area (19.1^0 N/ 84.8^0 E), and at Srikakulam, high dunes stabilized with vegetation and extensive saltpans can be seen near Naupada. Well defined beach ridges are seen at Uppada. Mudflat is also a striking feature in this area.

The AP coast like the entire east coast of India is predominated by depositional landforms such as beach ridge-swale complexes, mangrove swamps, mudflats, spits, barriers, lagoons, estuaries, and tidal inlets except in a few localities. Along the coast different types of coastal dunes are observed, which are mainly fore dunes and back dunes. As per the classification of Smith (1984) dunes are transverse, crescent shaped and parabolic. Along Visakhapatnam coast (17.7^0 N / 83.3^0 E) rocky headlands having a height of

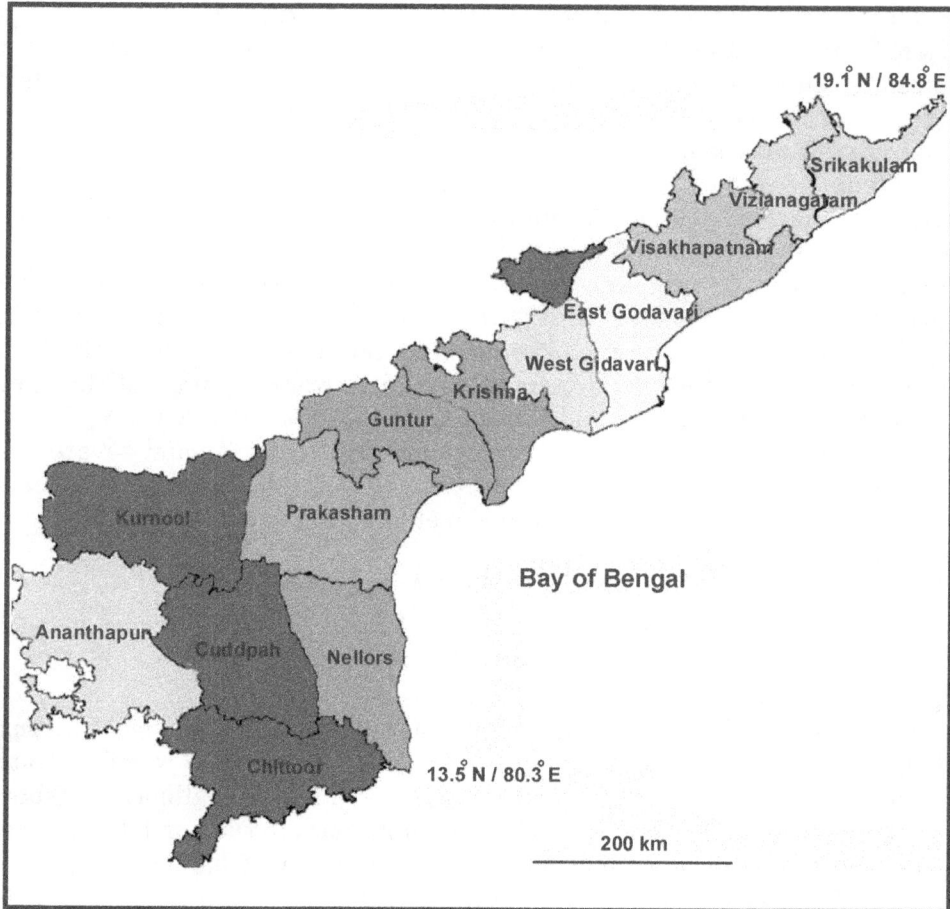

Fig. 63 : State of Andhra Pradesh

around 350 ASL can be seen fringed by 7 to 9 m high sea cliffs, narrow shore platforms and other related erosional landforms such as notches and caves.

Kikinada bay in the north, Krishna–Godavari twin deltas in the central parts, and the Penner delta and Pulicat Lake to the south along the AP coast are low-lying areas and exhibit landforms like beaches, mudflats, mangrove swamps, and tidal channels/creeks (Fig. 64) that enter up to more than 10–15 km inland even in this typical microtidal environment.

The Kakinada Bay : It is shallow (Depth 1m) at its southern portion and depth increases towards north to a maximum of 50m. Main area of sediment deposition is along the southern boundary of the bay where dense mangrove and mudflat development can be seen.

The bay near Nilarevu channel confluence has been nearly completely filled in last few years. The river mouth bar existing till 1975 has been transformed into spit resulting in the development of Kakinada bay and a lagoon. The southwest portion of bay at Gautami confluence is very shallow.

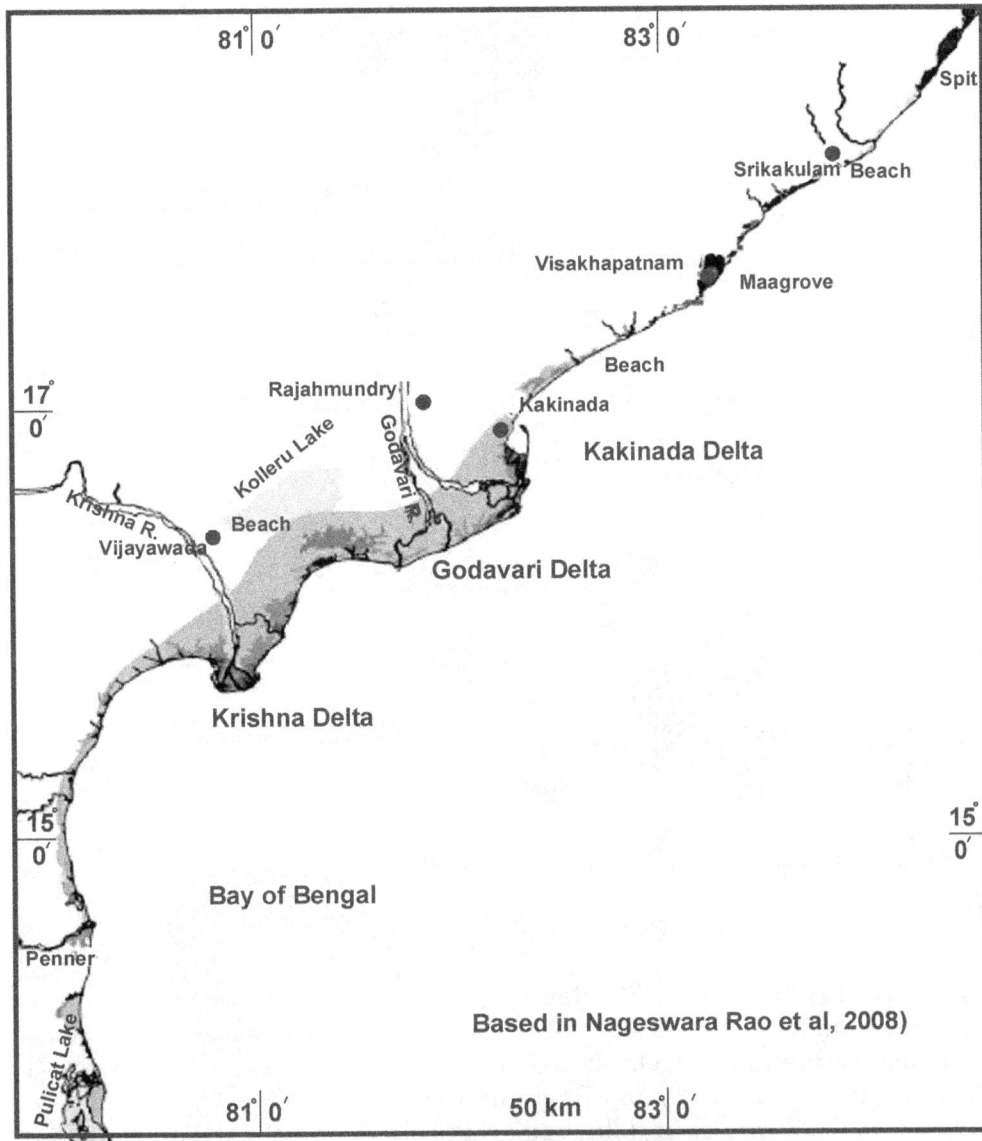

Fig. 64 : Coastal Features Of Andhra Pradesh

The region enjoys a meso tidal environment (2.2 m tidal range), The effect of tides is assumed to play a major role in spit formation in this delta. The Kakinada spit has reached its maturity state and is now getting eroded (Fig 65). It is thin all along its lower part and its width increases to 1.5 km towards the distal end. The spit now has a length of about 28 km. It has a curved head at the top. As there are no refractive waves within bay the sediments are getting accumulated at the distal end of bay to produce a wider head. The wide spit end attains maximum of 7 metres elevation above high tide water level. Head of the bay is occupied by dense mangroves. At the top, on both sides of the spit, wave cut terraces are being formed actively.

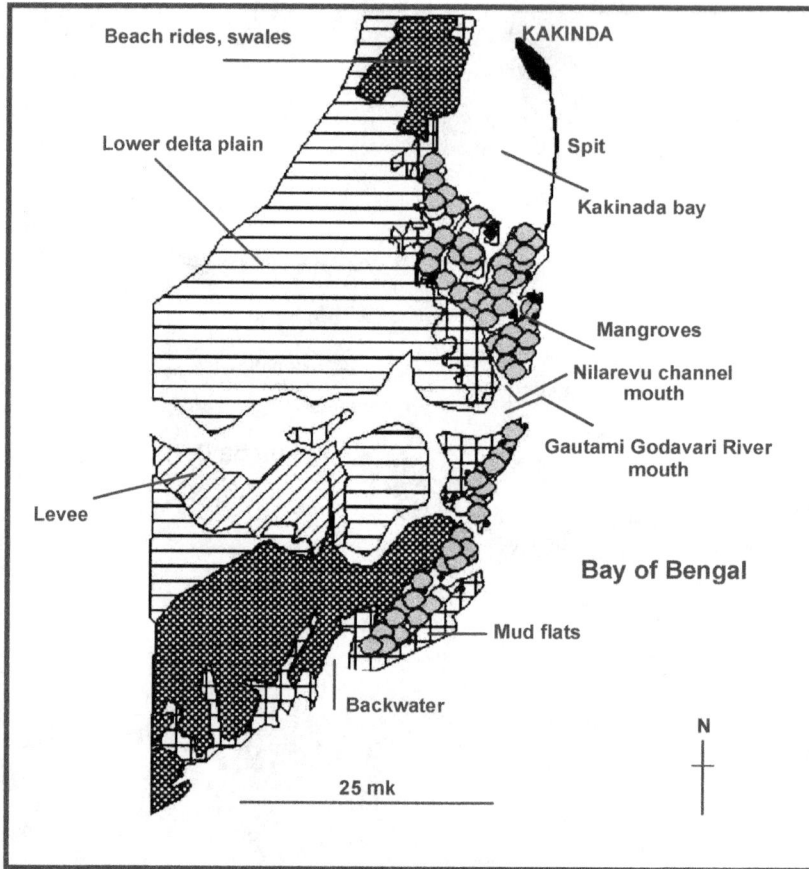

Fig. 65 : Kakinada spit

Krishna Godavari Delta : The delta region extends from Ongole(15.5⁰ N / 80.06⁰ E) to Machilipatnam, on both the sides of the river Krishna and upto Kakinada(16.9⁰ N / 82.2⁰ E) beyond Gautami river. On the north-western boundary lie Vijaywada(16.5⁰ N / 80.6⁰ E), Eluru(16.7⁰ N/ 81.1⁰ E) and Rajamundry (16.8⁰ N / 81.3⁰ E).

The Krishna-Godavari twin delta region is a lowlying plain with most part of it lying within 15 m above the mean sea level and gently sloping towards the Bay of Bengal. It covers an area of about 12,700 sq km. The length of the Krishna-Godavari delta-front shoreline is 370 km. (Fig 66). It faces a relatively low-energy marine environment with microtidal range of 1.5 m. The significant wave height is 2 m only (Nageswara Rao et al. 2008).

About seven km downstream of Rajahmundry city close to delta apex the Godavari River bifurcates into two distributaries – the Gautami and the Vasishta. Farther downstream, a third distributary, the Vainateyam branches out of the Vasishta, while the terminal branching of Gautami forms the fourth distributary mouth, the Nilarevu.

The Krishna River flows undivided for a distance of about 60 km from its delta apex near Vijayawada. There it branches out in a small distributary at Puligadda village.

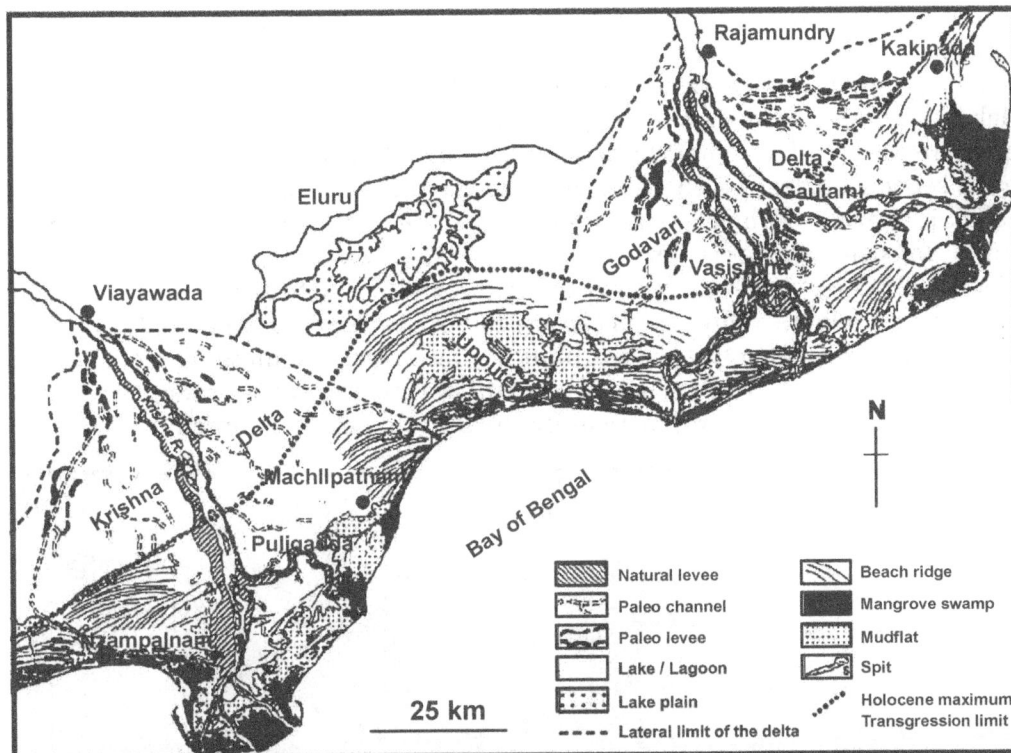

Fig. 66 : Krishna - Godacari Delta

Farther downstream, the river again splits into three distributaries just within 15 km from the shoreline forming the main delta lobe.

Morphologically, the entire Krishna Godavari delta plain appears as a single depositional unit. However, the two deltas are often separated by taking their respective limits along the lateral most abandoned distributary courses identifiable from the aerial photographs and satellite imagery (Sambasiva Rao & Vaidyanadhan 1979).

Nageswara Rao et al (2013) have given following detailed account of morphology of Krishna-Godavari twin delta region. There is a distinct difference in the morphometry of both the deltas. Comparatively, the Krishna delta is more elongated while the Godavari delta is wider. The long axis of the Krishna delta, when measured as a straight line distance from the delta apex at Vijayawada up to the most seaward part of the coastline is 95 km, while that of the Godavari from Rajahmundry is 75 km. But the length of the Krishna delta-front coastline is only 140 km against the 170-km-long Godavari delta front coast. The most prominent feature in the Godavari delta is the anomalous meandering of the distributary courses. Another significant feature of the Godavari delta is the remarkably straight segment of its delta-front coastline over a length of 70 km between the mouths of the Gautami and the Vasishta distributaries.

The Krishna River course from Vijayawada at the delta apex up to the shoreline is a 95 km long stretch. The Krishna delta shows a distinct bulge into the sea relative to the general shoreline trend on both sides of it.

The low lying inter delta coastal plain between the Krishna and Godavari deltas, is characterized by the presence of alternating lakes/ lagoons and series of beach ridges. The most prominent feature of this zone is the Kolleru Lake, a large freshwater body surrounded by much larger lake plain which is apparently the emerged part of the lake. The lake is located more than 35 km inland from the shoreline. A series of curvilinear beach ridges fringe the seaward margin of the Kolleru Lake. Apparently this lake was formed initially as a coastal lagoon by submergence of low-lying land when the shoreline was along the maximum Holocene transgression limit around 6.3–7.3 ka.

The Kolleru Lake has turned into a freshwater body due to diminished tidal influence, as the shoreline advanced by the progradation of the Krishna and Godavari deltas on both sides, and also due to freshwater inputs from several rivulets. The bed level of this lake in some parts is at or below sea level. Due to this seawater enters the lake during high tide through a 64 km long intricately meandering tidal channel.

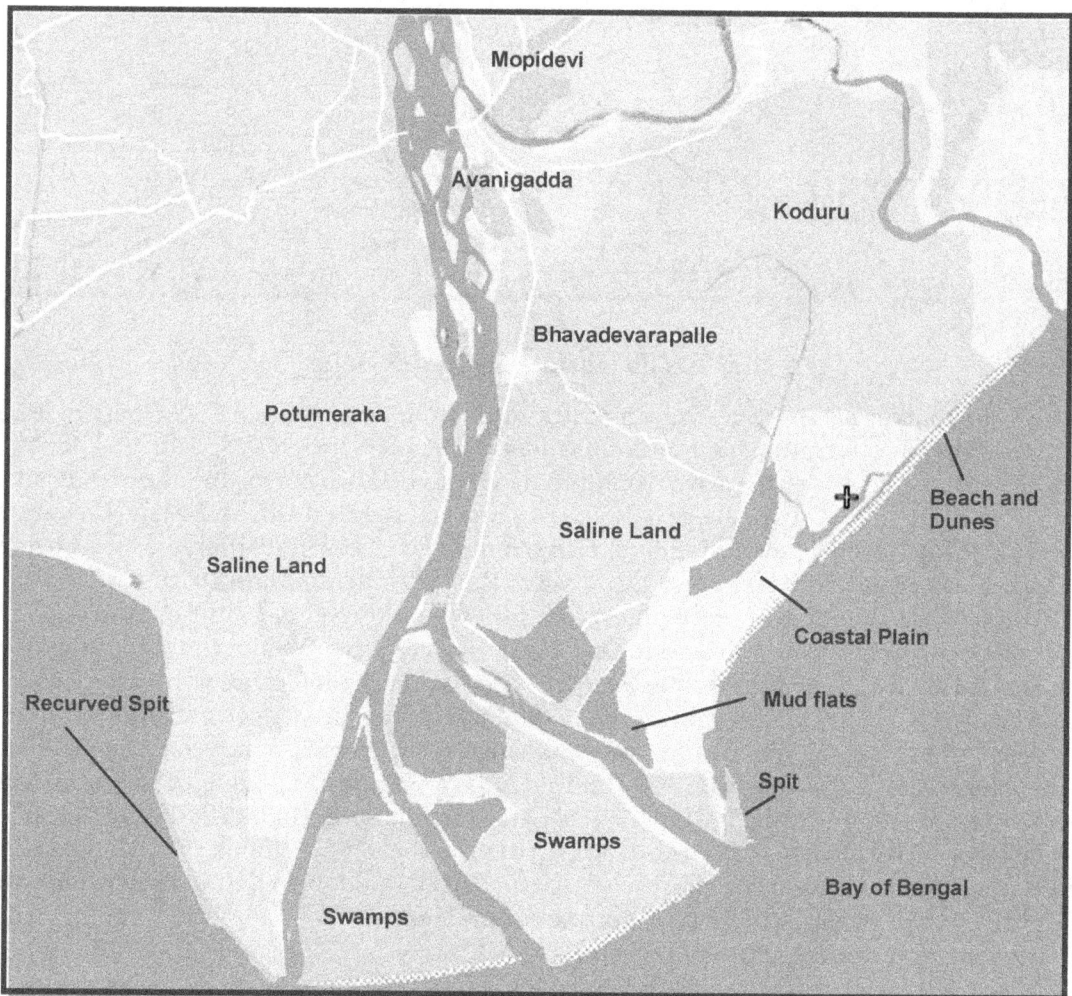

Fig. 67 : Krishna Delta

The topography in and around Kolleru Lake clearly shows that the lake area is a depression. It can be noted from the spot heights and benchmarks that the Kolleru Lake and its adjacent areas are within 2 m elevation surrounded by higher elevations in all directions ranging from 5 to 10 m. The lake area including the vast lake plain shows a general NE-SW linear trend.

Apart from the Kolleru Lake, there have been at least two more lagoons formed in the inter-delta zone of which Goguleru fringing the present shoreline is the active one. In between the Kolleru Lake (past coastal lagoon) and Goguleru (active lagoon), there is an extensive mudflat, which was thought to be a 'filled-in' lagoon (Nageswara Rao 1985b). The three lagoons are interspersed by two sets of beach ridges.

The present lagoon, viz. the Goguleru spreads along the inter delta shoreline and appears more like a linear body occupying low-lying swales in between beach ridges in the area. There are a number of sandy islands within the lagoon. The linearity and disposition of these islands indicate that they are the detached portions of the beach ridges that lie to the NW as well as SE of the lagoon.

Coastal and Deltaic Landforms in Krishna Delta :

Tidal amplitude in Krishna river delta is 1.5 m (Jain et al., 2010). The Krishna delta is much above the mean sea level as compared to the Godavari delta but nowhere the elevation is above 15m. The general slope of the delta is towards southeast.

The upper fluvial plain of the delta is characterized by the abandoned river courses and natural levees and the lower coastal strand plain includes beach ridges, mudflats, mangroves swamps, lagoon and spits which reflect the marine influence(Fig. 67). The other important coastal and deltaic landforms include beaches of deltafront sands, recurved spit which is conspicuously seen along the extreme right mouth of Krishna distributary showing upward and landward growth, Coastal dunes, Swamps and mudflats that are low lying areas occupied by muddy stagnant water. Areas on slightly higher mud flats usually support mangroves. On the left bank of Krishna and almost touching the Kolleru lake is an extensive coastal plain. Salt-affected lands in the delta are the leeward depressional areas, Relic lagoons occur as linear narrow strips in the coastal plain. Deltaic low lands occur as broad, winding, linear or irregular strips. The degraded deltaic plain is seen near the foothills or along the rivulets indicating degradation due to erosion.

Number of ancient beach ridges extend 35 km inland from the present coastline. These ridges date back to Holocene (Ranjana, *et al.*, 2015). Dense mangroves are seen around the distributaries in tidal creeks, channels, lagoons, tidal flats and mudflats.

The Pulicat (Pazhaverkadu) lake : is a brackish water lagoon and extends between 13⁰20' to 13⁰40' N latitudes and 80⁰14' to 80⁰15' E longitudes. It is formed out of backwaters of the Bay of Bengal and is the second largest brackish water lagoon in India having an area of approximately 600 sq km. The geomorphology reveals that water spread area of the lake is approximately 460 sq km. The Pulicat Lake shows diversified geomorphologic features including mud flats, sub littoral and littoral areas, sand dunes, sand spit bars, beaches, barrier islands, beach ridges and swale, sandy and muddy shores, saline, brackish and fresh water pools, and mangroves (Fig. 68). The Pulicate Lake has extensive tidal mudflats and a 41 km long spit beginning from Sriharikota.

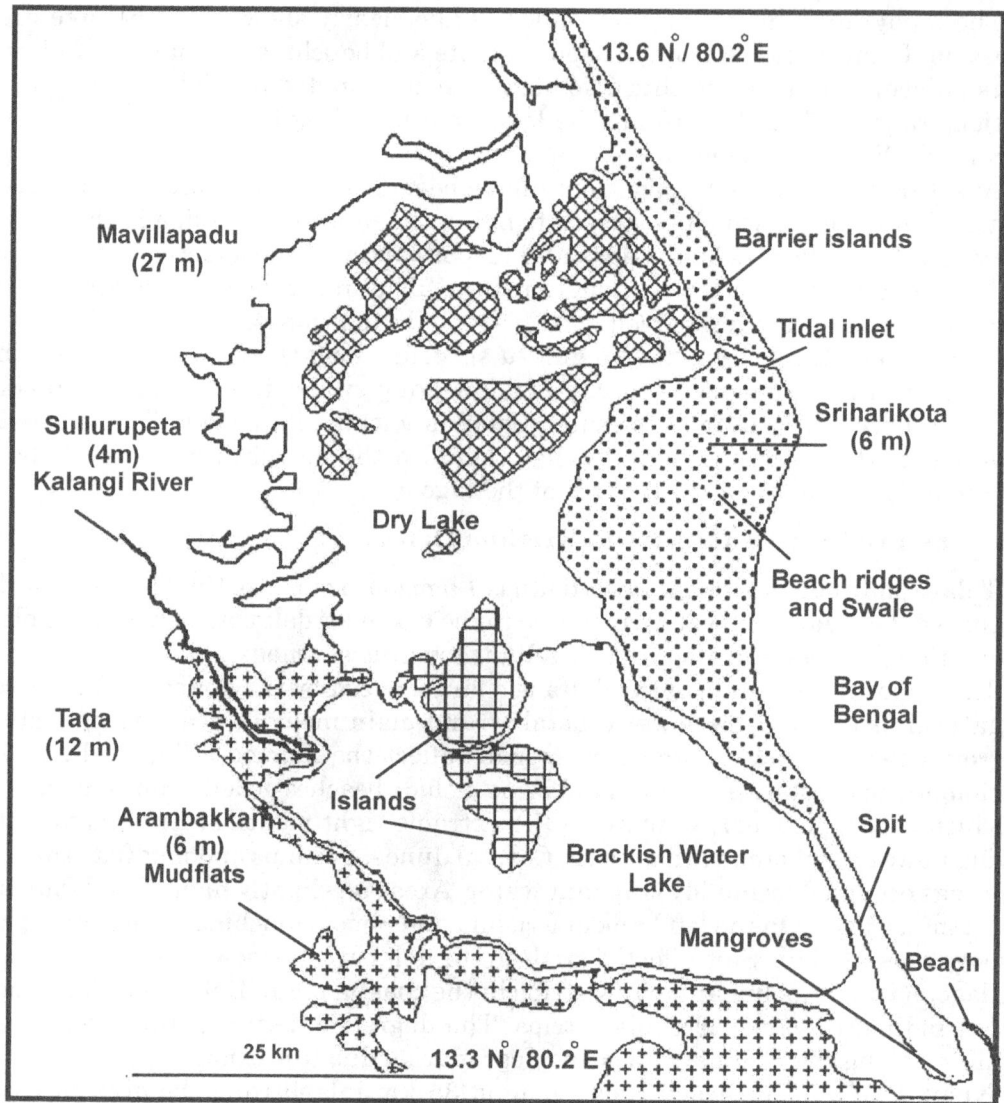

Fig. 68 : Features of Pulicat Lake

(C) The Coast of Odisha

The state of Odisha is located along northeastern coast of India between 17.8⁰ N and 22.6⁰ N latitudes and 81.4⁰ E and 86⁰ E longitudes. The 480 km long and 10 – 100 km wide coastline stretches from 19.1⁰ to 21.6⁰ north latitudes and 84.8⁰ to 87.5⁰ east longitudes. It covers six coastal districts, viz., Balasore, Bhadrak, Kendrapada, Jagatsinghpur, Puri, and Ganjam (Fig.69). The tidal sections of rives and their distributaries are confined to the lower reaches of the rivers in the coastal plain. The tidal sections show variations as per the shape of the river mouth, depth of the channel, and extension of the sand bars in the river mouths. The tidal channels vary from a maximum of 90 km in the case of Brahmani to a minimum of 5 km in the case of Baghuni from their respective mouths. The Mahanadi is tidal for about 35 km, whereas the Devi, a distributary of the Mahanadi is tidal for 45 km.

It is essentially a prograding and depositional coast, endowed with major estuaries. The Mahanadi, Brahamani and Baitarni rivers have formed large compound delta on the coastal region of Odisha. The river systems that carry the sediments to the coast are Subernekha, Bhudhabalang, Salanadi, Baitami, Brahmani, Mahanadi and Rushik1ya. These rivers carry a large volume of sediments that has produced a huge delta on the coast. The Orissa coast thus is under a continuous influence of freshwater flow from rivers and delta building processes on the coast.

It is gifted with Asia's largest brackish water lagoon, the Chilika; a 672 sq km extensive mangrove forest and wetland. It is observed that at the river mouths, there are a number of intermittent extensions of sand spits northward and repeated destruction of the same.

The majority of the geomorphic features along the coast comprise of sandy beaches, deltas, spits, mangroves, and mudflats and spread along 367 km long stretch. The stretch of coastline comprising estuaries and non mangrove vegetated coasts runs for about 74 km and about 39 km of coastline comprises of inundated coasts and cliffs especially along

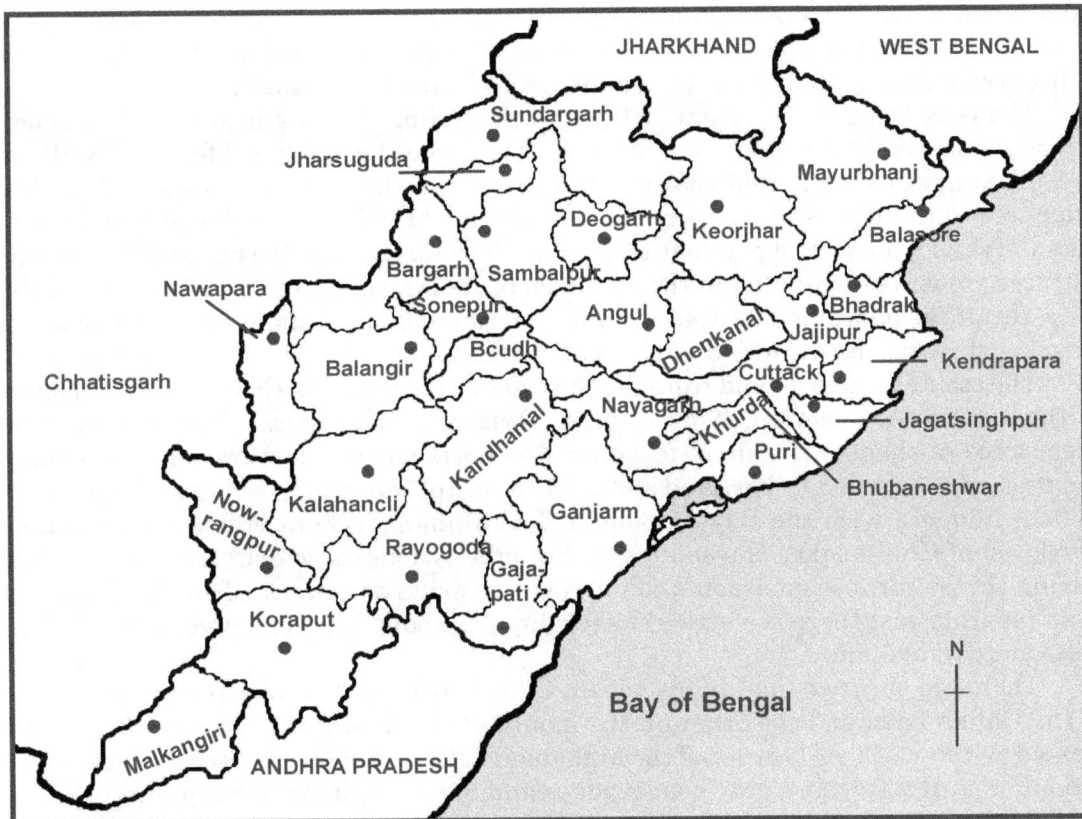

Fig. 69 : Sate of Odisha

the Chilika lake region (T. Srinivas kumar,2010).

The coastal districts of Orissa have experienced many major surges in the past. Extreme sea levels are a major cause of concern for coastal flooding in this region. The loss of land to the sea has now become a more recurrent phenomenon.

The mean significant wave height on this Coast ranges between 1.25 and 1.40 m. Tides along the Odisha coast are of mixed type and predominantly semidiurnal in nature. The average spring tidal range is 2.4 m and neap tidal range is about 0.9 m. About 302 km of coastline has a tidal range between 2.5 and 3.5 m along the northern coastal stretch from Puri (19.8⁰ N / 85. 8⁰ E), to Jagatsingpur, Kendrapada, Bhadrak, up to Balasore (22.4⁰ N / 87⁰ E). About 141 km of southern coastline has tidal range of less than 2.5 from Ganjam (19.3⁰ N / 85⁰ E) to Puri through Chilika Lake. The southern Puri has a tidal range of less than 2.5 m (T. Srinivas kumar,2010). Tidal range is less than 2 meters on the coast south of Puri.The Mahanadi River deltaic coast is micro-tidal with a mean tidal range of 1.29 m. The Odisha coast is wave dominated during monsoon season while it is mixed wave and tide dominated during non-monsoon period.

The ocean currents along this coast flow towards south with a speed ranging from 15 to 30 cm per second. On Odisha coast the mean significant wave height ranges between.1.25 and 1.40 m. During southwest monsoon high waves of height 9m or more are recorded which strike the coast obliquely and induce lirtoral long shore drift of 1.5 million cubic meter of sand annually from southwest to northeast in the near shore area (Ramesh et al, 2011). Along the east coast. longshore transport is southerly from November to February. northerly from April to September and variable in March and October.

The coast, in general aligns oblique to prevailing winds and waves that generate strong northerly littoral current. This coastal plain has been formed from older alluvium as well as newer alluvium of recent origin. The long shore sand transported in Odisha compares to the highest transport rates in the world, with the order of 1 million cubic meter of sand transported per year to north of Paradip (Govt of Orissa, 1974). The coastal features are wave dominated in the southern part of the coast (south of Mahanadi river). As a result, geomorphic features such as Barrier Island, spits, coastal lagoons, beach ridges and swales have formed in this part of the coast.

The coastal plains extend from the Subamarekha in the north east to the Rushikulya in the southwest. Coastal Odisha is characterized by wide deltas. The monsoons are a great force in shaping the shore features. The ports on the coast owe their existence to the projection afforded by bars and spits. The rivers of Odisha have created large deltas at their confluence with the Bay of Bengal. The Mahanadi delta starts its projection on northeast of Chilka lake. The sediments brought by longshore drift from the southwest during the Southwest monsoon and currents or drifts are arrested in the Chilka lake. Starting from east there is a straight shoreline for about 120 km between the Mahanadi delta and Srikakulam.

There are only two tidal inlets within these long stretches, one at the narrow mouth of the Chilka Lake and the other on the mouth of the Rushikulya River. Chilka Lake is located on the southwest corner of the Mahanadi delta and connected with the sea through a tidal inlet. It has wide sandy beach ridges and barrier spits which separate it from the Bay of Bengal.

The Odisha coast is a site of deposition formed and controlled by the Mahanadi and

Brahmani—Baitarani deltas. Mudflats, spits, bars, beach ridges, creeks, estuaries, lagoons, flood plains, paleo-mudflats, coastal dunes, salt pans, and paleo-channels are observed all along the Orissa coast. The Chilka lagoon is the largest natural water body of the Indian coast. The inlet mouth of Chilika lake is exposed to high annual north-ward littoral drift and observed to migrate about 500 m northward per year. The widths of beaches at Orissa also vary significantly. Littoral transport of sediments in the coastal region is a strong process. The coast is also exposed to severe cyclones. Turbidity in the nearshore as well as in the estuarine region is very high. Progradation of the coastal region in the north of the Devi estuary, and drifting of beaches has been observed. The Bhitarkanika and Hatmundia mangrove reserves are as extensive as 190 sq km. Gopalpur is rich in heavy minerals. Prominent and well-developed sand dune deposits containing monazite, zircon, rutile, ilmenite, and sillimanite occur along the southern coast of Orissa.

As mentioned earlier the coastal plain is drained by rivers such as the Mahanadi, Brahmani, Baitarni, Budhabalanga, Subarnarekha, Rushikulya and many other smaller streams. This region is called 'hexadeltaic region' of Odisha.It stretches all along the coast and has its maximum width near Mahanadi Delta. It is narrow around Balasore Plain and narrowest at Ganjam Plain. The North Coastal Plain comprises the deltas of the Subarnarekha and the Budhabalanga rivers and bears evidences of marine transgressions. The Middle Coastal Plain comprises of the compound deltas of the Baitarni, Brahmani and Mahanadi rivers and bears imprints of ancient coastal bays. The South Coastal Plain comprises of the lacustrine plain of Chilka lake and the smaller delta of the Rushikulya River.

Subarnarekha Delta :

The Subarnarekha deltaic barrier coast of Odisha is one of the most diverse, wave-tide mixed energy barrier coasts in India (Jana *et al*, 2014). The tidal range in the area is up to 3m. The wave height in fair weather season is 30 cm which increases to about 80 cm in south east monsoon. The deltaic coast exhibits features such as modern barriers, Holocene barriers, linear tidal basins, estuaries and estuarine islands.

The barrier spits developed along the delta coast protect the tidal basin of Subarnarekha delta from open marine environment (Fig. 70). The tidal inlets function as the drainage links between the lagoonal tidal basin and open marine region in the area. The morphology of spits, inlets and tidal channels inside the delta keeps on changing with tidal, riverine, longshore, storm related over wash deposition and windblown sand deposition in different seasons (Jana *et al*, 2014).

The growth of mangroves and salt marsh vegetation in the delta lagoon is mainly due to the presence of conducive substrate of silt clay deposits on the tidal flats in the delta. The western part of the tidal inlet is ebb dominated and shows development of significant ebb delta lobes. Linearity of the spit and shore parallelalism of offshore sand ridges suggests the dominance of long shore drift from west to east along the shore.

Mahanadi Delta :

The Mahanadi River delta lies between Puri (19^0 47' N / 85^0 47' E) and Kanakapur (20^0 35' N / 86^0 51' E). After traversing a distance of more than 800 km the Mahanadi River starts building up its delta plain from Naraj (20.47^0 N / 85.78^0 E) where the undivided

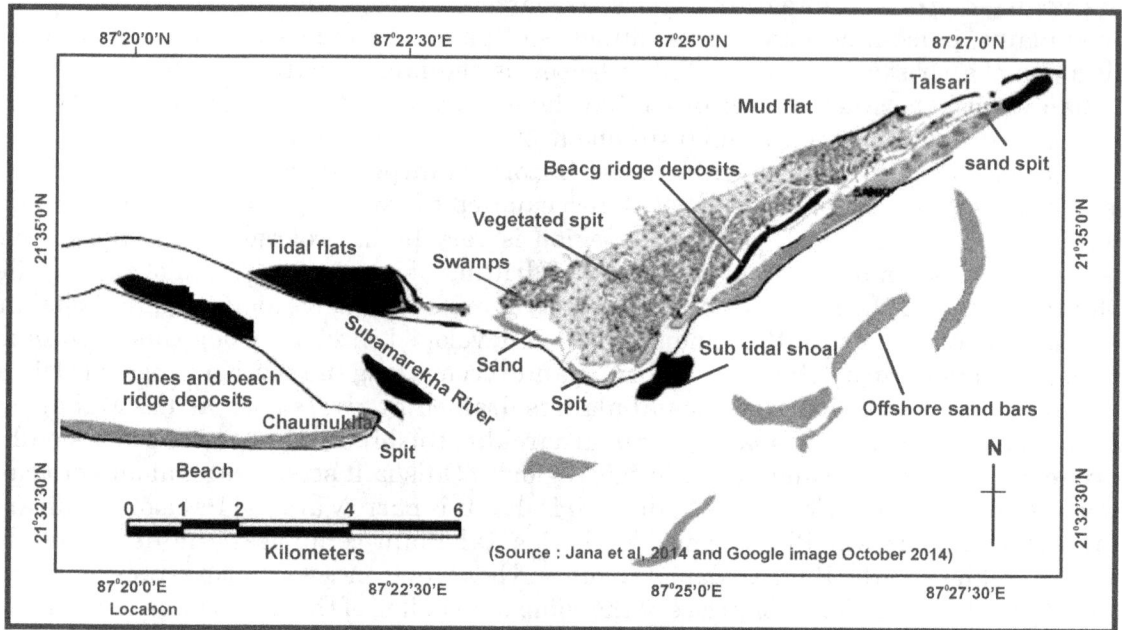

Fig. 70 : Subarnarekha Delta

Mahanadi branches forming its distributary system (Fig.71) ramifying in the delta plain area. Devi River is its principal distributary. Mahanadi River deltaic coast is microtidal with a mean tidal range measuring 1.29m. Tides are semi-diurnal. It is mainly a wave-dominated coast during the southwest monsoonal season, while during non-monsoonal fair weather season it is mixed wave and tidedominated region (Mohanti et al, 2005) Sediment deposition in the Mahanadi River deltaic environments is mainly monsoon-dominated.

The prominent features of the delta are ancient beach ridges. They are low sandy ridges lying parallel to the shoreline deposited by wave action(Fig. 72) from these beach ridges about three strandlines have been identified (Somanna, 2016). The Mahanadi delta is largely comprised of such beach ridges and ancient channels.

The beaches are backed by sand dunes.The width of the sand dune section varies from half a km to 2.5 km. The altitude of dunes varies from 0.5 meters to 3 meters. The dunes are normally covered with Spinifex type of spiny and bushy grasses. A prominent spit is formed north of confluence point of River Mahanadi. The spit has a length of about 25Km in north-south direction. The width of the spit varies from 200 meters to 500 meters. It looks like a hook for about 10Km from the confluence of River Mahanadi. The spit is formed by the deposition of sediments brought by littoral currents. Behind the spit a lagoon is formed. The spits are also noticed at the confluence of River Devi. These spits are relatively smaller in size.Like other modern deltas of the East coast of India the modern Mahanadi delta is presumed to be formed during Holocene period.

Tidal flats and mangrove swamps: Tidal flats and swamps are recognised all along the coast of the Mahanadi delta. Most of them, however, have a local extent around the mouths

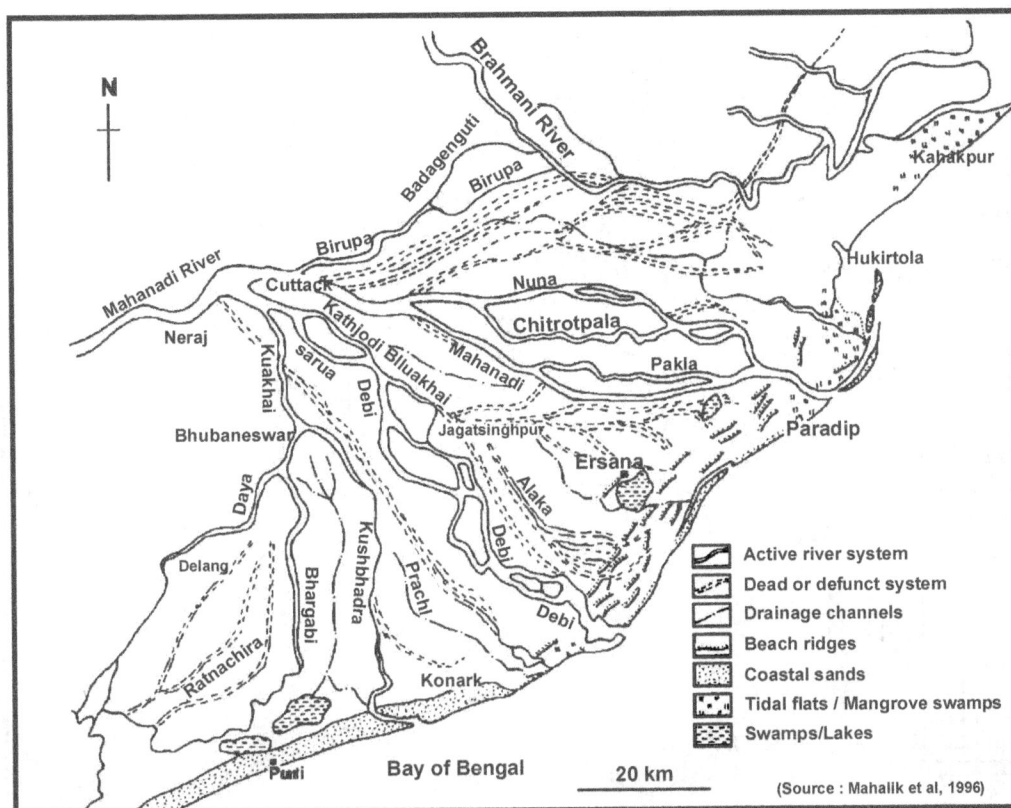

Fig. 71 : Mahanadi Delta

of distributary channels and in the swales between adjacent beach ridges (Mahalik, 1996).

Coastal sands and dunes: Widespread coastal sand bodies lie along the coast in the southern part of the Mahanadi delta, stretching from Chilika lake on the southwest to Konark on the northeast. They are as much as 15 m high and 2 to 5 km wide. These coastal sand bodies are made up of wind-blown sands covering the muddy deposits of a tidal flat or swamp origin. Parabolic dunes develop over them (Fig. 72).

Beaches and spits : The seaward margin of the delta plain is marked by a straight and continuously regular shoreline with a sandy beach all along it and without cliffs. Delta front environment supports a complex of barrier islands, bars and spit system. Accelerated longshore transport of sands helps building and prolongation of barrier island spit system. During high episodic monsoonal floods and storms damage is usually caused to sand spits and sand barriers.

Prominent spits occur near the mouths of the Mahanadi and the Debi rivers. The development and extension of spits towards the north are due to northerly moving littoral drift, which has constantly pushed the mouths of rivers towards the north. Thus the rivers Mahanadi, Debi, and Kushbhadra take a northerly course parallel to the shoreline for some distance. According to Manmohan Mohanti and Swain (2005) the delta has prograded in the seaward direction in the Holocene due to abundant supply of sediments from the

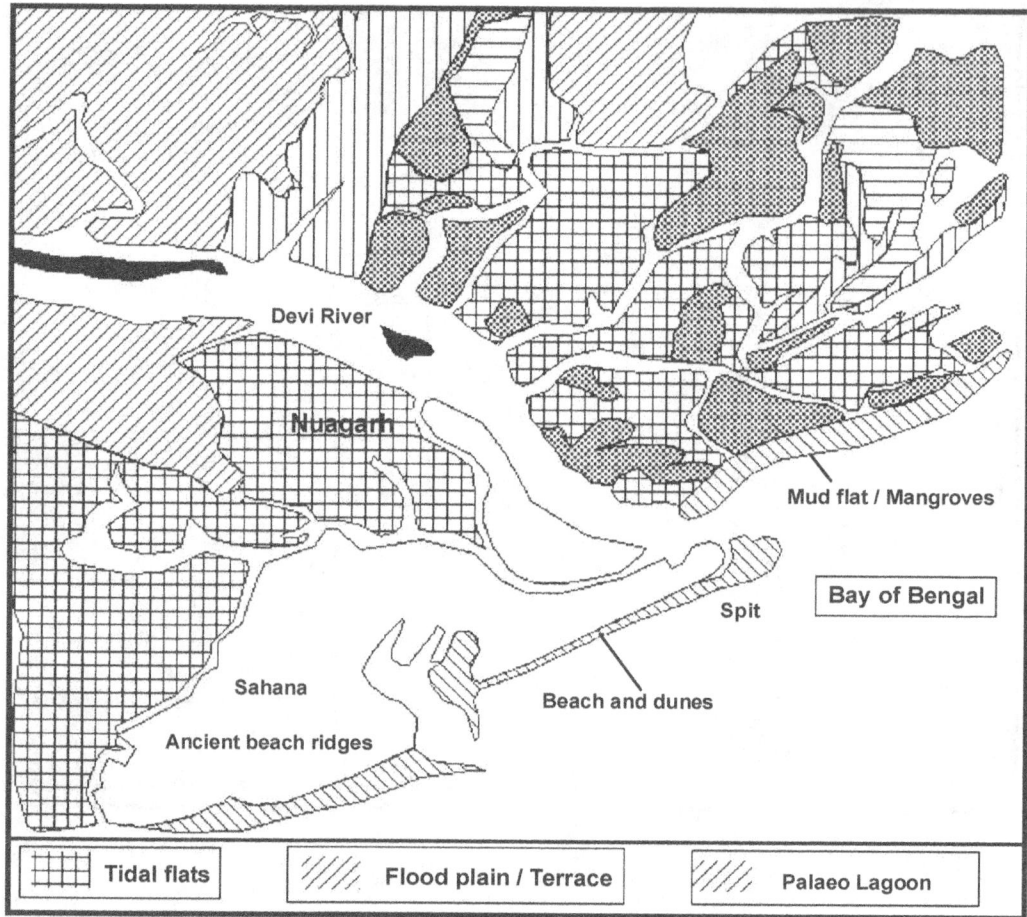

Fig. 72 : Features of Mahanadi Delta

hinterland under favourable climatic conditions, sea level adjustment, accommodation and tectonic subsidence.

The mangroves are located in the north and southern bank near Mahanadi river confluence. The marshes and mud flats are seen located near the shoreline and to the north and south of Devi river confluence also. The paleo lagoons and abandoned meander lobes have been identified by a Somanna(2016).

The Chilika Lake :

The Chilika Lake (lat. 19.47° to 19.9° N lat and 85.08⁰ to 85.63° E long) on the Odisha east coast is the largest partially enclosed brackish water lake in India. (Fig 73.). It is a shallow coastal water body separated from the Bay of Bengal by a long sand bar. The lake is a unique assemblage of marine, brackish and fresh water ecosystem with predominantly estuarine character. The pear shaped lake is about 60 km long and varies in width from 15 km in north to 5.5 km in south. The average water spread area of the lake is 900 sq km in

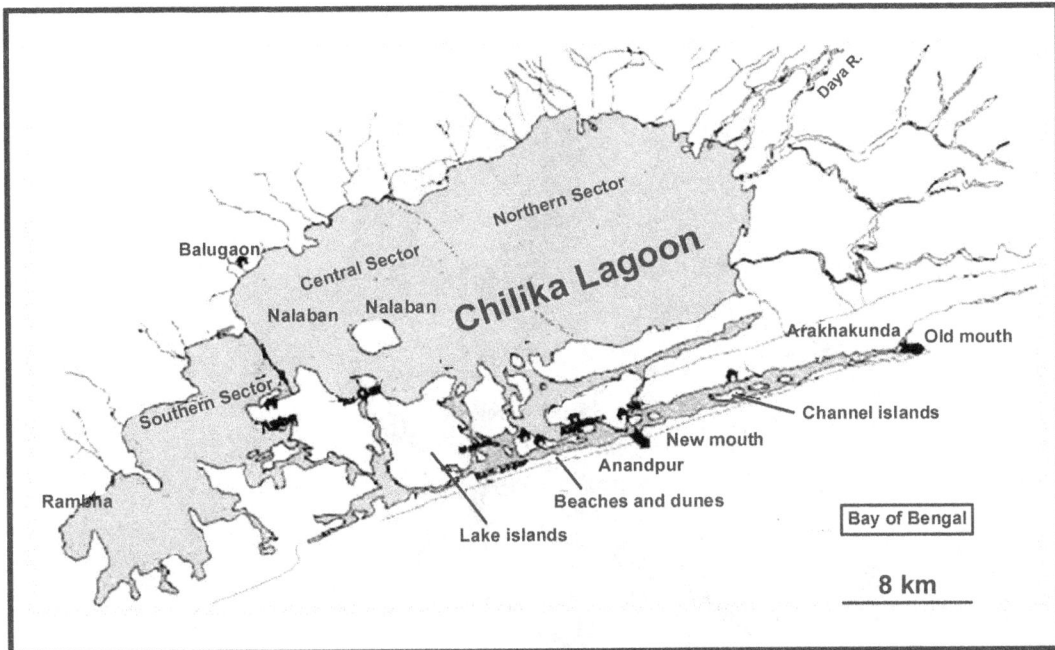

Fig. 73 : Chilka Lake

dry season (December-June) and 1160 sq km in the wet season (July-October). The water depth in the lake varies from 0.9 to 2.6 m in dry season and 1.8 to 5.7 m in the wet season.

Topographically the Chilka Lake can be divided into sectors depending on salinity and depth such as the Northern, Central and Southern sectors and the Outer channel connecting the lake to sea at Anandpur (19.7° N / 85.5° E). Due to its rich biodiversity it was designated as a "Ramsar Site". It consists of 65 sq km of mangrove swamps and other wetlands and 90 sq km of mud flats covered by grass.

The study of shoreline change by Ministry of environment and forests (Ramesh et al, 2011) shows that the coast of Odisha is largely accreting (46.8%) and 14.4% is stable. Erosion (High. medium and low erosion) accounts for 36.8% of the coast. The accretion is dominant in the central and northern part while erosion dominates the southern part of the coast. Low to medium erosion is found to occur along 137 km of total 480km coast of Odisha.(Fig.74).

(D) The Coast of West Bengal

The location, of the West Bengal state is very strategic. It borders Odisha in the southwest and Bihar & Jharkhand to the west. It shares a long border with Bangladesh on the east and in the north, shares borders with Assam, Bhutan and Nepal. The southern boundary of the state is formed by Bay of Bengal. There are 19 districts in West Bengal out of these only three districts East Medinipur, North 24-Parganas and South 24 Parganas fall within the coastal zone (Figure 75).

The length of the coastline in West Bengal is 220 km and it stretches between 21.6°

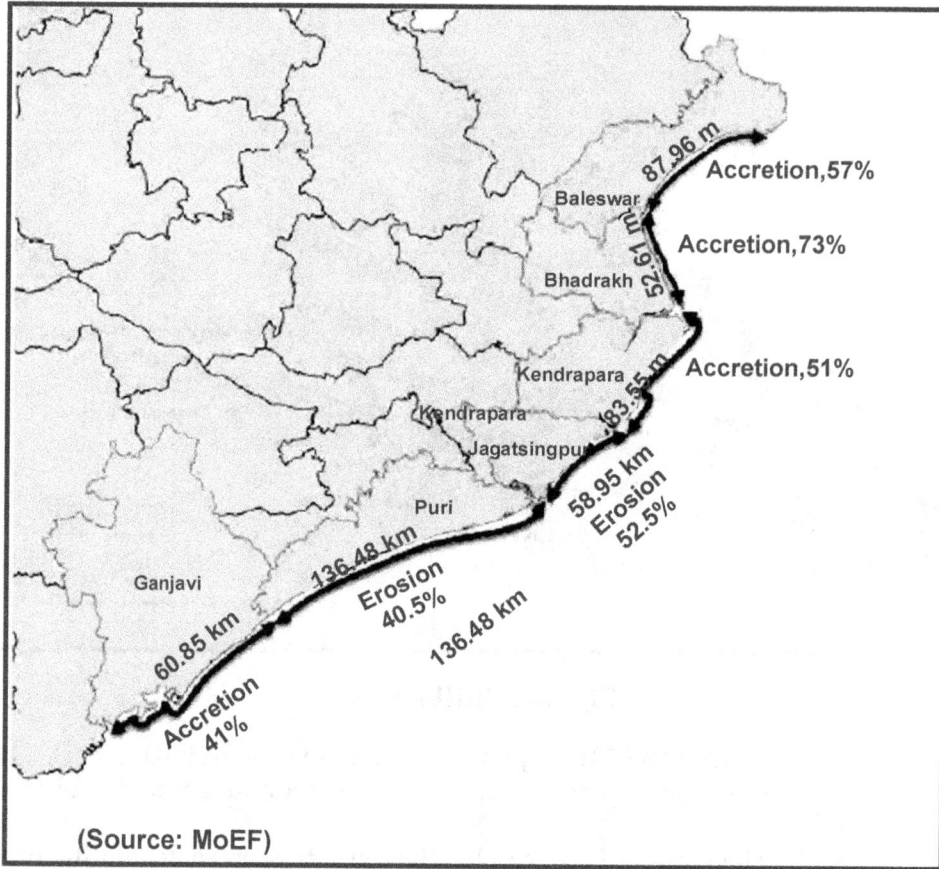

Fig. 74 : Erosion and Accretion on Odisha Coast

and 21.5⁰, N latitudes and 87.5⁰ and 89.9⁰, E longitudes. Based on tidal amplitude the coast can be sub-divided into two different coastal environments namely (1) Hugli estuarine plain characterized by a network of creeks encompassing the islands with dense mangrove vegetation and offshore linear tidal shoals from Sagar Island (21.6⁰ N / 88.1⁰ E) to the border of Bangladesh to the east. This zone enjoys a macro tidal environment where tidal range is greater than 4 m and (2) Medinipur (Digha-Sankarpur-Junput) coastal plain to the west of the Hugli estuary with rows of sandy dunes separated by clayey tidal flats from Sagar Island to Orissa border to the west where tidal range varies from 2 to 4 m (Govt WB,2010) making it a mesotidal environment (Fig. 76). In the coastal area of West Bengal a clay blanket of 20 to 30m thickness is generally present below which brackish water aquifers occur within 120 m depth in the western part of Hugli river and within 150 to 180 m in the eastern part of it (Govt of WB,2010).

Medinipur Plain :

Within the coastal region of Medinipur Plain, water-logged marshy areas, like the Dankuni, Ghatal and Dubda are characteristic back swamp areas of the landscape.The Kanthi coastal plain lying between the estuary of the Subarnarekha River and Hugli River

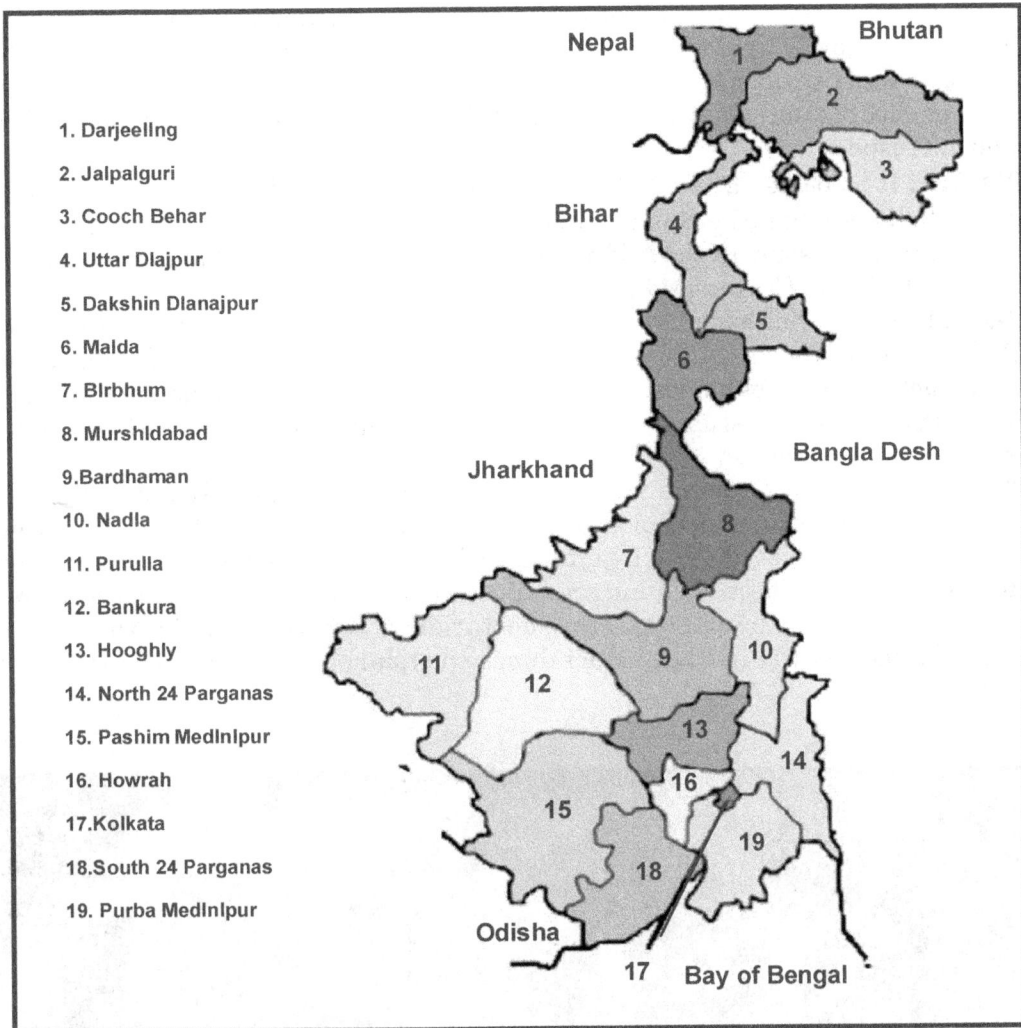

Fig. 75 : State of West Bengal

1. Darjeellng
2. Jalpalguri
3. Cooch Behar
4. Uttar Dlajpur
5. Dakshin Dlanajpur
6. Malda
7. Blrbhum
8. Murshldabad
9. Bardhaman
10. Nadla
11. Purulla
12. Bankura
13. Hooghly
14. North 24 Parganas
15. Pashim Medlnlpur
16. Howrah
17. Kolkata
18. South 24 Parganas
19. Purba Medlnlpur

is a northeast-southwest trending coastal zone characterized by rows of beach ridges and intervening low-lying swales (mudflats). Being a deltaic low lying coastal stretch this area is monotonously flat alluvium surface having a height ranging between 5 and 7 m above MSL. This part of the coast is a meso-tidal, wave dominated coast. The wave environment of the Kanthi coast is dominated by wind driven waves coming from SE or SSE. Wave height is below 1 m which increases significantly towards the near shore area. The predominant direction of littoral drift is from west to east although a mild littoral drift from east to west during winter months has been recorded (Kunte *et al*, 2001). The shoreline between Digha and the Rasulpur is characterised by sand dunes and a wide sandy beaches (Fig.77). Several small tidal creeks cross the shore, the largest being the Shankapur Creek east of Digha. The dunes at Digha are presently getting eroded at a very rapid rate.

Midnapore coast is characterised by sand dunes, long shore currents, high salinity, low turbidity and low vegetative cover.A 60 km long coast extends from the west bank of Hugli Estuary to New Digha at the extreme south west (longitudinal extension 87'20E to 88'5'E and latitudinal extension 21'30'N to 22°2'N). Although very short in length, it displays unique occurrence of dunes, mudflats, sandflats. The mean tidal range at Digha and Haldia is 4.2 and 4.9 respectively.

Talsari : It is located at 21.6^0 N lat. and 87.4^0 E. Long. Talasari is one of the less exploited Odisha beaches. The Subarnarekha estuarine delta is the westernmost unit of the topographic expression in the Midnapur coastal plain. A tract of mangrove swamp is preserved at the back of Talsari tidal and intertidal flats.

Digha Beach : The beach has a linear almost east-west extension of variable width from Digha creek in the east to Jan Nala in the west near the mouth of Subarnarekha river. The backshore of Digha coast is about 100m wide with multiple small dunes. At many places on the beach estuarine mud mixed with sand is seen (Das,2015). The beach sediments are highly compact

The continuous row of dunes from Subarnarekha estuary to Kanthi exhibits the ancient dunes which were formed during Holocene period. The shoreline shifted further southward with the regressive phases of the sea and the ancient sand dunes are seen slightly inland from the oresent coast. These old dune are reduced in height. Around Digha there are two major dunes developed by aeolian transportation of sands from the the foreshore (Das,2015). All the features like bars, older dunes and tidal basin are parallel to the present day coastline.

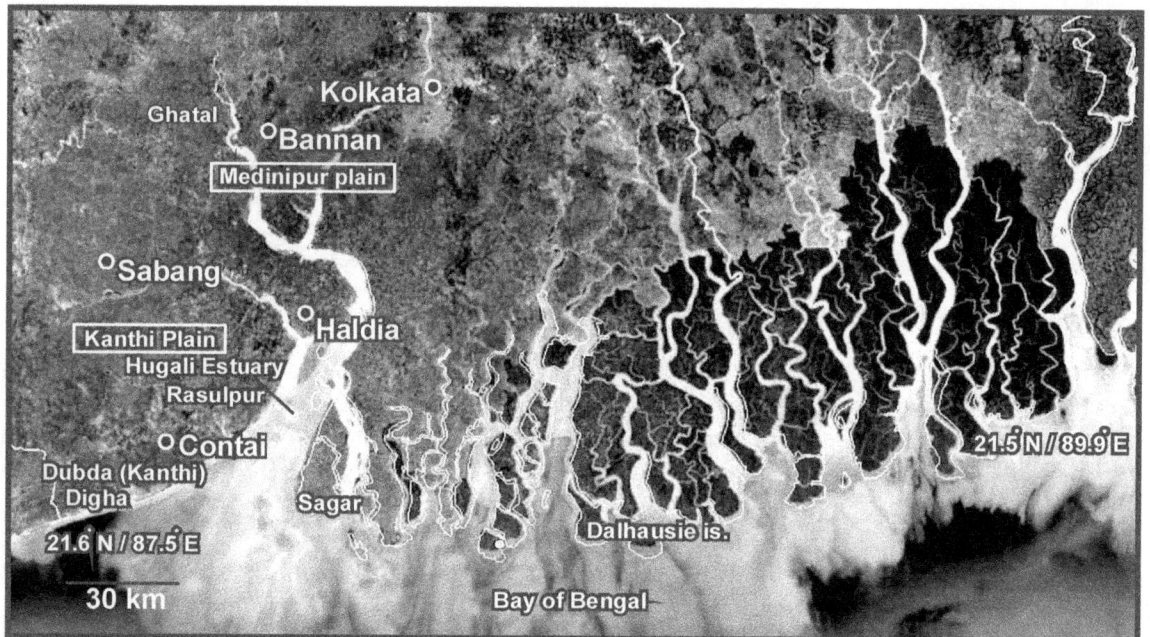

Fig. 76 : Features of West Bengal

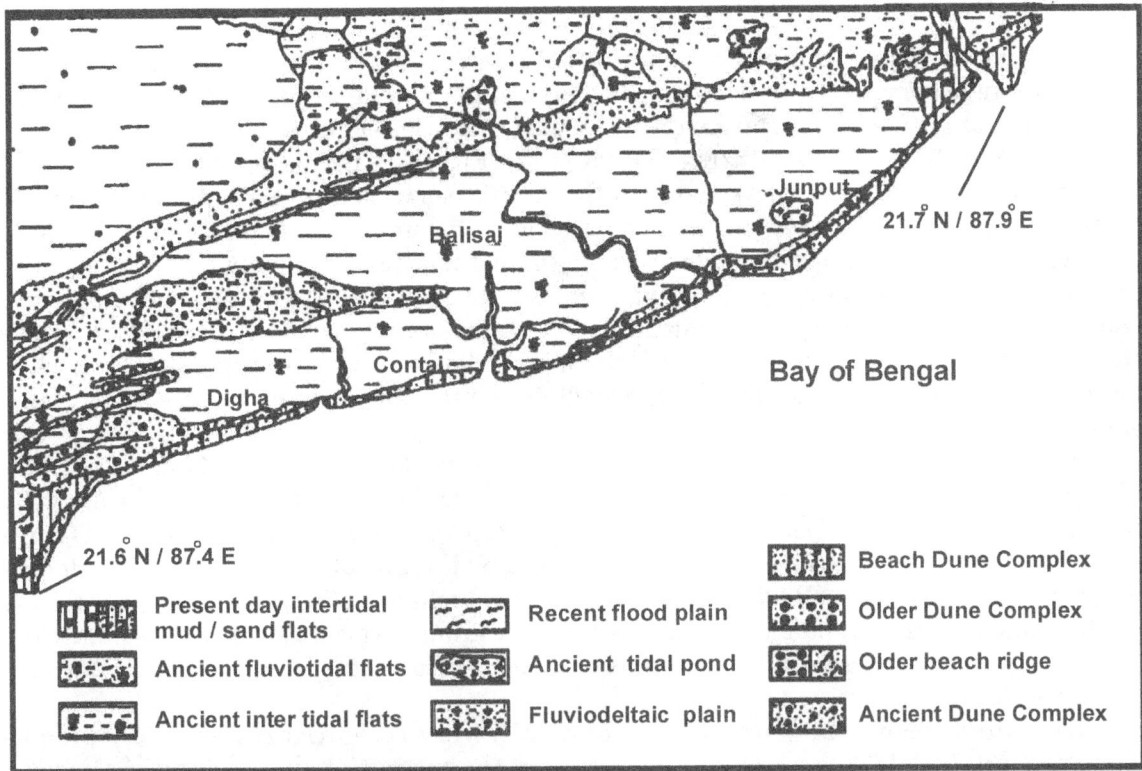

Fig. 77 : Features of Medinipur Plain

Junput : Junput located to the north east is characterised by 1.5 km wide intertidal mud flat having intricate network of tidal chaanels and a recent patch of mangroves comprising of Rhizophora and Avicennia.It is fronted by a 5 km long sand spit bar.

Nayachar Island: It is a mid creek island (21.9^0 N / 88.1^0 E) in the macro- tidal estuary of Hugli River. It has been recently raised considerably by continuous accretion of sediments since 1945 (Paul et al, 2004). It is colonised by thick growth of salt marshegrass and mangrove swamps. Between 1967 and 1997, the island progressively enlarged, yielding an average accretion rate of 0.88 sq. km per year (Hazra and Sanyal, 1996).

Beaches : The beaches along Midnapore coast are associated with beach berm, beach face, ridge and runnel, rip channels, low tide terrace, long shore trough, long shore bar and significant back wash ripples within the limits of back shore, foreshore, inshore and adjacent shallow water area of offshore zones (Das,2015). The beach face is marked by sedimentary structures and active bioturbation. The fine sand beaches of Midnapore coast have gentle foreshore slopes The spit bars such as near Serpurjalpai (21.6^0 N / 87.7^0 E) and Junput (21.7^0 N / 87.8^0 E) are usually developed along drift alignment beaches due to movement of long shore sediment. The extensive tidal flats of the foreshore region are made up of silt and clay sediments.

Hugali Delta and Plain :

The Ganga (Hugli) delta in the south 24 Parganas district is a tide dominated delta. The tidal range sometimes exceeds 5m. The area supports a dense mangrove forest in the inter-distributary marshes of large funnel-shaped estuaries opening to the sea southwards with wide mudflats exposed during the low tide (Fig.78). The mangrove plants help to build the land by their decaying material that combines with the fine sediments trapped by the mangrove roots. The entire landscape is made up of islands separated by tidal channels. The delta is apparently formed by amalgamation of such islands. Narrow silty to sandy beaches could be seen along the sea. The off-shore areas of the delta are characterized by coast perpendicular tidal shoals and channels up to the shelf edge. Near the coast, some tidal shoals are exposed during the low tide and are rapidly colonized by mangrove plants. The shoals gradually grow in size by trapping of sediments and organic debris.

Sundarbans : The Indian part of Sundarbans lies between 21°30' N and 22°40'48" N latitude and 88°1'48"E and 89°04'48" E longitude (Fig.79). It is delimited in the north by the northern extension of the intertidal zone marked by mangrove forests of 1830 (Hazra et al, 2002).In the south, the Sundarbans is bound by the Bay of Bengal. The river Hoogly (in the west) and the river Harinbhanga– Raimangal –Ichamati (in the east) demarcate the western and eastern boundaries respectively. The Indian part consists of 9630 sq. km area and out of the 9630 sq. km. 2195 sq. km. is claimed by Wetland – Mangroves and 2069 sq. km. by of tidal rivers.

The Sundarban island system is geologically very recent(Hazra et al, 2002). The Delta outbuilding of Ganga-Brahmaputra system though initiated at the end of Miocene, could have reached the present location of Sundarban delta, not more than 10,000 years back (Pleistocene to Recent). During the recent times the delta acquired a typical tide dominated lobate form with a tidal range varying between 3.7 to 5 m. The flood and ebb tides have a semi-diurnal nature (12.5 hrs interval) and occur twice daily. Within this cycle floodwater flows for 2-3 hrs duration. In the remaining 8-9 hrs, the area is covered by ebb tide flow of lesser velocity. The tide dominated estuarine system exhibits typical flow separation, with downstream freshwater flows along the right bank and upstream saline water tide flowing along the left bank of the channel.

There are many prominent islands along the Sundarban coast facing the sea. The islands of 3 to 8 m height,are partially/ often completely inundated by water during high tide,. The subsidence in these areas is comparatively more severe than in the open coastal parts of Digha. From the very begining, Sundarban with it's numerous large and small islands remains a subsiding delta, with prominent vertical upbuilding.

The south western corners of the sundarban islands are particularly susceptible to sustained erosion (Govt of WB,2010). According to Govt. of West Bengal report (2010) the western banks of the inner islands are more vulnerable to erosion than the eastern banks and the rate of retreat of western banks is more severe. Accretion is localized in the inner estuaries particularly along eastern and northern margins and along the coasts of islands trending parallel to the incoming waves.

The mangroves on the West Bengal coast are mainly colonized in this Sundarbans area, which is the largest single block of tidal halophytic mangroves (1430 sq km) in the world. The rivers mostly carry the freshwater from the upper reaches. The dense mangrove

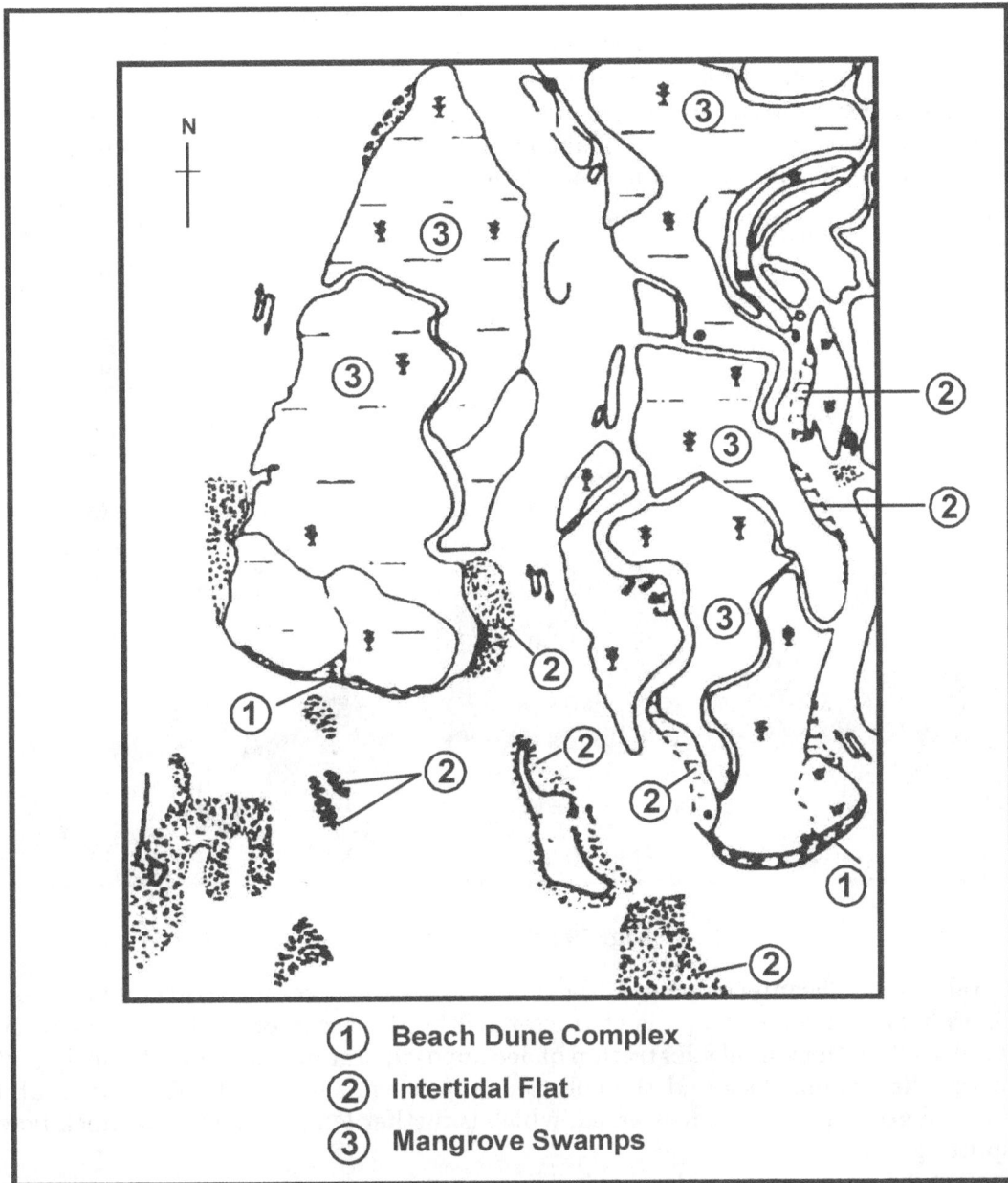

Fig. 78 : Mouth of Hugali Creek

forest of Sundarban is famous for the dangerous man-eating Royal Bengal Tiger and the most ferocious crocodiles.

The major geomorphic features in the delta region are mudflats, bars, shoals, beach ridges, estuaries, and a network of creeks, paleo-mudflats, coastal dunes, islands and salt pans.

In such an estuarine or coastal plain areas, the surficial deposits and the geomorphology fail to provide a chronological/chronostratigraphical account of the development of the thick pile of sediments (Parthasarathi,1995).

The Sunderbans Mudflats are found at the estuaries and on the deltaic islands where there is a low velocity of river and tidal currents. The flats are exposed at low tides and submerged in high tides, and thus the unstable mudflat changes morphologically even

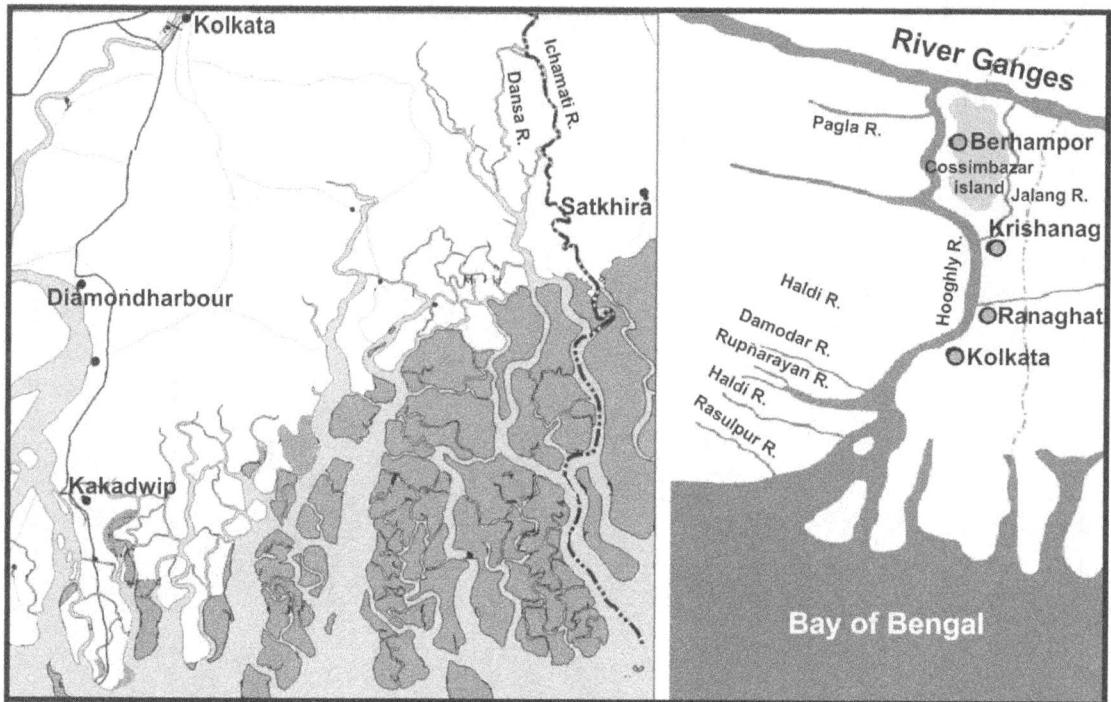

Fig. 79 : Sundarbans

in one tidal cycle. The interior parts of the mudflats are the home for luxuriant mangroves. The morphology of the swamps is characterized by the occurrence of saltpans, ditches and banks with a thick mud substratum of decomposed organic matters. The spring tides submerge the swamp floor and the ebb tides affect the slope of the floor with lateral erosion and gradually form a new creek, which is further lengthened by the quick flow of the splitting tides.

Coastal dune systems in the Sunderbans comprise a system of low ridges parallel to the coast, separated by large dry and wet sand flats. The coastal dunes of the western islands are now being engulfed by the encroaching sea (Bose, 2002). The rate of coastal erosion is high on the coast of Fraserganj, Bakkhali and Sagar. The main rivers of this region, including the estuarine and all having a southward course towards the sea are the Hugli, Piyali – Bidyadhari, Muri Ganga, Saptamukhi, Thakuran, Matla, Gosaba and Harinbhanga. They are separated from one another by numerous islands and the important ones from west to east are Sagar, Fraserganj, Lothian, Bulcherry and Halliday Island. Between the large estuaries and rivers, there are innumerable watercourses called "Khals" (canals) forming a perfect network of channels, drawing water from every block of land. Each block is like a saucer with high ground and one or more depressions and is drained off by the hierarchy of surrounding creeks(Bose,2002).

Sea level rise : Sea-level rise is the greatest threat and challenge to West Bengal coast. A 45 cm rise in global sea levels would lead to the destruction of 75 percent of the Sundarban mangroves. Along with global sea level rise, there is a continuous natural subsidence in the Sundarban, causing a rise of about 2.2 mm per year. The resulting net rise rate is estimated at 3.1 mm per year at Sagar. There are many consequences of this SL rise (Govt of WB,2010) in terms of flooding of low-lying deltas, retreat of shorelines, salinization and acidification of soils, and also the changes in the water table.

Significant beach lowering has been observed over the erosional domains of the coastal tract. It is seen that the erosion/submergence dominates in the southern part of the island system. This points to a possibility of relative rise in sea level in this part of the Bay of Bengal, rather than paucity of sediment supply. This is also confirmed from the open Digha– Junput coastal stretch which shows a parallel retreat of shore line (Hazra et al, 2002).

REFERENCES

- **Agarwal D. P. and Guzder S. J. (1972) :** Quaternary studies on the western coast of India, Preliminary observations, The palaebotanist 21, pp. 216-222.
- **Ahmed (1972) :** Coastal Geomorphology of India, Orient Longman, New Delhi, p. 222.
- **Alcock A. (1992) :** A naturalist in Indian seas; or four years with the Royal Indian Marine Survey Ship "Investigator". (John Murray, London).
- **Allen J. R. L. (1984) :** Sedimentary structures, their character and physical basis, in Developments in Sediment logy, Elsevier, New York.
- **Allersma E. (1976) :** Holocene tidal sedimentation. Ed. Klein G. D. V., Benchmark papers in Geol. Vol. 30, Hutchinson Ross, Pennsylvania.
- **Allersma E. (1982) :** Mud in Estuaries and along coast. Chinese society of hydraulic engineering, China, Publ. 270.
- **Anand V. K., Venkataraman G., Vishwanathan S., Rao G. S. K. (1987) :** Some aspects of Geomorphic processes along the Goa coast through remote sensing, in Jr. of Ind. Soc. of remote sensing 15(1), pp 49-54.
- **Angusamy N., Geetha S. and Rajamanickam, G. V. (1992) :** Beach placer minerals exploration along the coast between Mandapam and Kanyakumari, DOD proj. rep. (Unpublished), p 50.
- **Anil Babu, Satya Narayan P., Guru Prasad (2015) :** Study of Geoinformatics for East coast of India along the Andhra Pradesh Coast, India, in Int. Jr. of innovative research in Sci., engineering and tech. Vol. 4 (II).
- **Anitha Shyam and G. Sreenath (2006) :** Ground Water Information Booklet of Lakshadweep Islands, Union Territory of Lakshadweep
- **Antony M. K. (1976) :** Wave refraction off Calangute beach, Goa, with spl. ref. to sediment transport and rip currents, in Ind. Jr. of Marine Sci. Vol. 5, pp 1 -8.
- **Avinash Kumar, Narayana A. C., Jayappa K. S. (2010) :** Shoreline changes and morphology of spits along southern Karnataka, west coast of India: A remote sensing and statistics based approach, in Geomorphology Vol. 120, pp133 -152.
- **Bandyopadhyay, A. (1989) :** Geol. Surv. India, Spec. Publ., 24, 343–347.
- **Banerjee and Sen (1987) :** J Earth Sci 4.
- **Barua, K. D., Steven, A. K., Richard, L. M. and Williard, S. M. (1994) :** *Mar. Geol.* 120, 41-61.
- **Baskaran M., Marathe A. R., Rajaguru S. N., SomayajuluB. L. K. (1986) :** Jr. of Archaeological sciences vol. 13, pp 505-514.
- **Bhatt Nilesh and Bhonde Uday (2006) :** Geomorphic expression of late quaternary sea level changes along southern Saurashtra coast, western india, in Jr of Earth science systems 115(4), pp 395-402.

- **Bhatt Nishith, Majethiya heman V., Solanki Paras m. (2016)** : Geomorphology of coast between Koteshwar and Jakhau (Kori Creek), Kachchh, Western india, in Int. jr. for sciietific research and development 4 (6), pp 2321-0613.

- **Bhattacharyya P., Nayak B., Singh R., Maulik S. C, (2008)** : Studies on beach placers of Kerala coast

- **Bird, E. C. F. (1985** : *Coastline Changes: A Global Review*, John Wiley, UK, p. 219.

- **Bose Shivashish (2002)** : The Sundarbans Biosphere : A study on uncertainties and impacts in active delta region, Centre for built env. Kolkata.

- **Boothroyd V. C. (1978)** : Mesotidal inlets and Estuaries in Coastal Sedimentary Environments, Ed. Davis R. A., Springer Verlag New York.

- **Bruckner H. (1987)** : New data on evolution of Konkan, western India, Explorations in Tropics Ed Datye et al K. R. Dikshit felicitation volume

- **Bruckner, H. (1988)** : Indicators for formerly high sea levels along the east coast of India and on the Andaman Islands. Hamburger Geographische Studien, 44, 47-72.

- **Bruckner H. (1989)** : Late Quatenary Shorelines in India In : Scott, pirazzoli and Honig (Eds) Late Qnantermary Sea-level correlation and applications (Eds D B Scott et al.)

- **Carter R. W. G. (1989)** : Coastal Environments, Academic press, London, pp 335-375

- **Chakraborty Susanta Kumar (2006)** : Coastal environment of Midnapore, west Bengal: potential threats and management, in Jr. of Coastal env. Vol 1 (1)

- **Chandramohan, P. and Nayak, B. U. (1994)** : J. Coast. Res., 10, 909–918.

- **Chandramohan, P. (1988)** : Longshore sediment transport model with parti-cular reference to Indian coast. Ph D thesis, IIT Madras, p. 210.

- **Chandramohan, P., Jena, B. K. and Kumar, V. S. (2001)** : Littoral drift sources and sinks along the Indian coast. *Curr. Sci.*, 81, 292-297.

- **Chandramohan, P., Kumar, V. S. and Nayak, B. U. (1991)** : Wave statistics around the Indian coast based on ship observed data. *Indian J. Mar. Sci.,* 20, 87-92.

- **Chandramohan, P., Kumar, V. S. and Nayak, B. U. (1993)** : Coastal processes along the shorefront of Chilka lake. Indian J. Mar. Sci., 22, 268-272.

- **Chandramohan, P., Kumar, V. S., Nayak, B. U. and Pathak, K. C. (1993)** : Variation of longshore current and sediment transport along the south Maharashtra coast, west coast of India. Indian J. Mar. Sci., 22, 115–118.

- **Chandramohan, P., Nayak, B. U. and Raju, V. S. (1998)** : Coast. Eng., 12, 285–298.

- **Chandramohan, P., Raju, N. S. N., Kumar, V. S., Anand, N. M. and Nayak, B. U. (1991)** : Coastal processes along the south Karnataka coast. Technical Report NIO/TR/15-91, National Institute of Oceanography, Goa.

- **Chandrasekar N., saravanan S., Rajamanickam M., Rajamanickam G. Victor (2014)** : The spatial variability of ridge and runnel beach morphology due to beach placer mining along Vembar – Kallar coast, India, in Annales, Polonia, Vol. LXIX 2, pp 61 -62.

- **Chauhan. O. S. (1989)** : Sedimentilogical parametrs of beach sediments of the east coast of India. Journal of Coastal Research, 6, 573-585.

- **Chauhan, O. S. and Vora, K. H. (1990)** : Continent. Shelf coast of India, Proc. Recent Res. Geol. W. India, *Res.,* 10, 385–396.

- **Cooke, R. U. and Doornkamp, J. C. (1974)** : Geomorphology in Environmental Management, Clarendon press, Oxford.

- **Das Nirmalya (2015)** : Dune changes and its impact at Digha coast of West Bengal, in Int. Jr. of Soc. Sci. and interdisciplinary research, Vol. 4 (4), pp 38 - 40

- **Davies J. R. L. (1977)** : Geographical variation in coastal development. Longman, London pp 123-128.

- **Davis R. J. (1978)** : Coastal Sedimentary Environment Springer Verlag, New york.

- **Desai Kasturi N. & Untawale Arvind G. (2002)** : Sand dune vegetation of Goa: Conservation and management, Botanical society of India

- **Deshmukh Sanjay (1989)** : Nature's coast guard The wwf quarterly, Bombay pp 7 to 9.

- **Deswandikar A. and Karlekar S. N. (1996)** : A geomorphic significance of coastal dune complex at Diveagar, Ind. Jr. of Geomorphology, Vol. 1 & 2 Academic and Law serial. New Delhi, pp 125-135.

- **Deswandikar, A. K. (1993)** : Nearshore Sediments, Sedimentary features and Tidal basins alongh Diveagar - Aravi - Valvati coast Maharashtra, Unpublished Ph. D. Thesis, University of Pune pp. 60-101.

- **Dhanunjaya Rao, G., Krishnaian Setty and Raminaidu, C. H. (1989)** : Heavy mineral content and textural characteristics of coastal sands in the Krishna- Godaveri, Gosthani-Champavati, and Penna river deltas of Andhara Pradesh, India: a comparative study. Exploration and Research for atomic minerals. Atomic minerals division, Hyderabad Vol. 2, pp 147-155.

- **Dhavalikar, L. N. (2002)** : Sea level scenario on Vengurla coast of Maharashtra, Transactions. Institute of Indian Geographers, Vol. 24, No. 1 & 2, pp 63-71.

- **Dickison K. A., Barry Hill H. L., Holmes C. W. (1972)** : Criteria For Recognising Ancient Barrier Coastlines. Recognition of Ancient Sedimentary Environments, Ed. Ragbyn and Hamblin, Tulsa (pp 192-214).

- **Dikshit K. R. (1970)** : Geomorphology of Gujarat, National Book Trust, New Delhi 260 P.

- **Dikshit K. R. (1975)** : Geomorphic features of the west coast of India between Bombay and Goa, Geographical review of India Vol. 38 PP 260 –281.

- **Dikshit K. R. (1986)** : Mahatashtra in maps, Maharashtra state board for literature and culture, Mumbai,

- **Dutta S. (1991)** : Exploration of ilmenite resources in the bay off Ratnagiri coast; Past and Present. Marine wing Newsletter, Geol. Survey of India Vol. VII, 1, pp 5-6.

- **Dyer (1979)** : Estuarine hydrography and sedimentation Cambridge, London

- **Fairbridge R. W. (1968)** : 'Thc Encyclopedia of Geomorphology', (Dowdcn, Hutchinson & Ross, Inc.)

- **Faruque B. M., Ramachandran K. V. (2014)** : Continental shelf of Western India, Geol. Soc. of London Memoires 41: pp 213 - 220

- **Fischer H. B., List J. E., Koh R. C. R., Imberger J., Brooks N. H. (1979)** : Mixing in inland and coastal waters, Academic Press, New York.

- **Ganapathi S. and Pandey A. N. (1991)** : Evolution of Landforms on Narmada and Tapi Deltas, Gujrat, in Quaternary Deltas of India, Ed Vaidyanadhan R. Geol. Soc. of India, Bangalore, Memoir 22.

- **GardinarJ. S. (1903)** : Introduction, In : The Fauna and Geography of the Maldive and Laceadive Archipelogoes, (Ed. J. S. Gardiner), pp. I-II, (Cambridge University Press, Cambridge).

- **Ghatpande Jyoti (1993)** : A Lithified Eolian Coastal Sand Dune at Kelshi, Mahashtra. Coastal Geomorphology of Konkan Ed. Karlekar S. N. Aparna Publications, Pune (pp 182-195).

- **Gilson, J. L. (1959)** : Sand deposits of titanium minerals. Mining Engineering, vol. 214 pp 421-429.

- **Ginsburg R. G. (1976)** : Tidal deposits, a case book of recent examples and fossil counterparts, Springer Verlag, New York.

- **Gopinathan C. K., Qasim S. Z. (1974)** : Mud banks of Kerala – Their formation and characteristics, in Ind. Jr. of Mar. Sci. Vol., 3 pp 19.

- **Govt. of West Bengal (2010)** : Integrated coastal zone management of West Bengal coast, state project report. Inst. of Env. Studies and wetland management, Kolkata, pp 8-11.

- **Greenwood B. & Davis R. A. (1984)** : Hydrodynamics and sedimentation in wave dominated coastal environments, Dev. In Sedimentology Vol. 39, Elsevier, Oxford.

- **Gujar A. R. (1995)** : Morphogenetic controls on the distribution of the littoral placers along central west coast of India, Proc. Recent Res. Geol. W. India, Oct. 29 – 31, pp 171-1803.

- **Gujar A. R. (1996)** : Heavy mineral placers in the nearshore areas of South Konkan, Maharashtra: Their nature, distribution origin and economic evaluation, Ph. D. thesis, Unpubl., Tamil university, Thanjavur, P 234.

- **Gupta G. V. M (2002)** : Geographical information for Gulf of Kuchchh; Govt. of India, Dept. of Ocean development, ICMAM, Chennai, p 53.

- **Gupta, S. K. (1972)** : Chronology of the raised beaches and inland coral reefs of Saurashtra coast, Journal Geology, 7: 1-7.

- **Gupta Vishal and Ansari A. A. (2014)** : Geomorphic portrait of the little Rann of Kutchchh, in Arabian jr of Geosciences, 7 (2), pp 527-536.

- **Gupta S K and Amin B S (1974)** : Io/U ages of corals from Saurashtra coast; Marine Geol. 16 79–83.

- **Guptha, M. V. S. and Hashimi, N. H. (1985)** : Indian J. Mar. Sci., 1985, 14, 66–70.

- **Guzder (1975)** : Quaternary environments and Stone Age cultures of Konkan, Coastal Maharashtra, India, Deccan college Pune

- **Hails J. and Carr A. (1975)** : Nearshore sediment dynamics and sedimentation. John Wiley and Sons, London.

- **Hanamgond P. T. (2007)** : Morphodynamics of the beaches between Redi and Vengurla, Maharashtra, Tech. Report DST Govt of India.

- **Hanamgond P. T. (2012)** : Esturies : dynamics of Kalavali, Kolamb and Karli rivers, with spl. ref. to impct on Malvan coast, VGC major research proj. 268 p.

- **Hanamgond P. T. (2015)** : Esturies : A review of literature on defination and classification, in Abs. vol. Not semi on Estuaries of India, Past, present and future.

- **Hashimi, N. H., Nair, R. R. and Kidwai, R. M. (1978)** : Indian J. Mar. Sci., 7, 1–7.

- **Hashimi, N. H. (1992)** : Correlation between Offshore Sediments and Coastal Features, In, Workshop on Coastal Geomorphology 14-16 Nov 1990. Andhra University Visakhapatnam sponsored by DST New Delhi. pp. 37-45.

- **Hashimi, N. H., Nigam, R., Nair, R. R. and Rajgopalan, G. (1995)** : Holocene sea-level fluctuations on Western Indian Continental margin: An update, Jr. Geol. Soc. India 46, pp. 157-162.

- **Hazra and Sanyal A:(1996)** : Ecology of collembolan in a periodically inundated newly emerged alluvial island in the river Hoogly, West Bengal. Proc. Zoology Society, Calcutta, 49(2). pp. 157-169.

- **Hazra and Sugata, Ghosh Tubin, Das Gupta Rajashree, Sen Gautam : (2002)** : Sea level and associated changes in the Sundarbans, in Sci. and Cult. Vol. 68, pp 9-12.

- **Hegde A. V., Akshya B. J. (2015)** : Shoreline transformation study of Karnataka coast-Geospatial approach: in Aquatia procedia Vol 4, pp 151-156.

- **Hesp P. A. (2000)** : The beach backshore and beyond, in Short A. D. 'Beach and Shore face morphodynamics' School of Geosciences, University of Sydney, pp 48-76.

- **Hiranandani, M. G. and Ghotankar, S. T. (1961)** : Technical Memoran-dum, NAV2, Central Water and Power Research Station, Poona, pp. 1-42.

- **Indian Tide Tables (2006):** Indian and selected foreign ports, Govern-ment of India, New Delhi, p. 231.

- **Jain, I., Rao, A. D. Jitendra, V. and Dube. S. K., (2010)** : Computation of Expected Total Water Levels along the East Coast of India, in Journal of Coastal Research 26 (4), 681-687.

- **Jana Subrata, Paul Ashis Kr., Sk Majharul Islam (2014)** : Morphodynamics of Barrier Spits and Tidal Inlets of Subarnarekha Delta: a study at Talsari-Subarnapur spit, Odisha, India, in Ind. Jr. of Geography and Environment, (13), pp 23-32.

- **Jayappa K. S., Mitra D., Mishra A. K. (2006)** : Coastal geomorphological and land use and land cover study of Sagar island, Bay of Bengal, India, using remotely sensed data, in Int. Jr. of Remote sensing, 27 (17), pp 3671-3682.

- **Jelgerma, S. (1966)** : Sea level changes during the last 10, 000 years. J. S. Sawyer (ed) World climate from 8, 000 years B. C to O. B. C. London Metrological Society, pp. 44-69.

- **Jelgersma, S. (1971)** : Sea Level Changes During the last 10, 000 years, in An introduction to coastline development, Ed. Steers J. A., MacMillan, London. pp 24-47.

- **K. Nageswara Rao, P. Subraelu, K. Ch. V. Nagakumar, G. Demudu, B. Hema Malini, A. S. Rajawat and Ajai (2013)** : Geomorphological implications of the basement structure in the Krishna-Godavari deltas, India, in Zeitschrift für Geomorphologie Vol. 57, 1, 25-44, 2013 Article published online April 2012.

- **Kumaran K. P. N., Limaye Ruta B., PadmalalD. (2012)** : India's fragile coast with special reference to Late quaternary environmental dynamics, in Proc. Ind. Nat. Sci. Acad. 78(3), pp 344-345.

- **K. P. N. Kumaran, Ruta B. Limaye, Sachin A. Punekar, S. N. Rajaguru, S. V. Joshi, S. N. Karlekar (2012)** : Vegetation response to South Asian Monsoon variations in Konkan, western India During the Late Quaternary: Evidence from fluvio-lacustrine archives, Quaternary International xxx (2012) pp 1to16.

- **Kale V. S. and Awasthi, Anita (1993)** : Morphology and formation of armored mud balls on Revadanda beach, Western India, Jr. of Sed. Petrology, 63 (5) pp 809-813.

- **Kale Vishwas and Rajaguru, S. N. (1985)** : Neogene and Quaternary trangressional and regressional history of west coast of India-An overview, Deccan College bulletin, Pune pp 153-163.

- **Kaliasundaram, G., Govindasamy, S. and Ganesan, R. (1991):** Coastal erosions and accretions. In *Coastal Zone Management (In Tamil Nadu State, India)*, (eds Natarajan, R., Dwivedi, S. N. and Rama-chandran, S.), Ocean Data Centre, Anna University, Chennai, pp. 73-82.

- **Kar, A. (1993)** : Neotectonic Influences on the Morphological Variation along the Coastline of Kutch, India Geomorphology Amsterdam Vol 8 No 2 and 3 pp 199-219.

- **Kar Amal, Abichandani R. K., Anantharam, K., Joshi, D. C. (1992)** : Perspectives on the Thar and the Karakum, Dept. of Science and Technology, Ministry of Sc. And Tech., Govt. of India, New Delhi.

- **Karlekar S. N., Jog S. R. (1982)** : The planation surfaces and the development of drainage of Maharashtra plateau, in Inst. Ind. Geographers 4(1), pp 91-98

- **Karlekar Shrikant (1982)** : The coastal dunes and the dune building plants of Kalbadevi spit bar, Maharashtra, in Inst. Ind. Geographers 5(1), pp 73-76

- **Karlekar Shrikant, Dikshit K. R. (1983):** Formation, thickness and composition of Konkan laterites, in The National Geog. Jr. of India, 29 (3 and 4), Calcutta, pp 164-175.

- **Karlekar Shrikant (1985)** : The tidal landforms of Uran, Alibag and Murud coast of Maharashtra, in trans Inst. Ind. Geographers 7(2), pp 157-161.

- ————————————— **(1986)** : Fossil mangroves and coastal change at Revas, in National Geographer 30 (1 and 2) Allahabad, pp 155-162.

- ————————————— **(1986)** : Holocene sea level fluctuations along Diveagar coast, in National Geographer 30 (1 and 2) Allahabad, pp171-177.

- ————————————— **(1990)** : The significance of Harihareshwar shore platforms in the interpretation of sea level changes along Maharashtra coast, Trans. Inst. Ind. Geographers 12(2l), Pune, pp 139-143.

- ————————————— **(1990)** : Reef lagoon, and beach sediments of Kayaratti coral island, Lakshadweep, India, in Ann. Nat. Ass. Geog. 10 (2) New Delhi, pp 59- 61

- ————————————— **(1993)** : Coastal geomorphology of Konkan, Ed, Shrikant Karlekar, Aparna publ Pune, p 325.

- ————————————— **(1993)** : Shifting sediment concentration and sedimentation in Banganga estuary on the Arabian sea coast, India, Trans. Inst. Ind Geographers 15(2l), Pune, pp 45-52.

- ————————————— **(1994)** : The Geomorphic significance of marker horizons in fluvio marine deposits of Konkan estuaries, MBP research jr, Pune 8 (21) pp1-6

- ————————————— **(1995)** : Movement of silt and clay through ebb current in

Dudh river estuary Palghar, in Indian Geomorphology Vol 1, Ed, Jog S. R., Rawat Publ., New Delhi pp 181-188

- ——————————————— (1995) : A study of tidal clays in the southern arm of Banganga River estuary, Indian Geomorphology, Vol 1, Ed Jog S. R., Rawat Publ. New Delhi pp 189-194.

- ——————————————— (1995) : Sea walls on Maharashtra coast : an assessment, MAEER'S MIT Pune Jr., 3 (12), Pune, pp 57-59.

- **Karlekar Shrikant, Devne Manoj (1995) :** The occurrence of mud on a sandy pocket beach near Anjarle, Maharashtra, trans. Inst. Ind. Geographer 17(1), pp 7-14.

- **Karlekar Shrikant, (1996) :** Sediment control of Dharamatar and Dabhol creek, MAEER'S MIT Pune Jr. spl issue on Coastal Environmental management 4 (15 and 16) Pune, 93-98.

- **Karlekar Shrikant, Dhavalikar Lakshmi (1996) :** Coastal Environmental management in India–An Overview, Spl issue on Coastal Environmental management 4 (15 and 16) Pune, 11-13.

- **Karlekar Shri kant, Deswandikar A. K. (1996) :** Nearshore environments of Dive agar coast, coauthor A. K. Deswandikar, in MAEER'S MIT Pune Jr. spl issue on Coastal Environmental management 4 (15 and 16) Pune, 35-46.

- **Karlekar Shrikant, (1996) :** The hydrodynamics and sedimentation of Konkan estuaries, Geog Review of India Calcutta 58 (1), pp 27-32.

- **Karlekar Shrikant, Deswandikar A. K. (1996) :** A geomorphic significance of coastal dune complex at Dive agar, coauthor Abhay Deswandikar, Ind. Jr. Of Geomorphology 1 (2) Academic and Law Serials New Delhi pp 125-135.

- **Karlekar Shrikant, (1997) :** Shore normal monsoonal dynamics of a sandy beach at Kashid, coastal Maharashtra, Deccan Geographer, 35(2) Pune pp 111-120.

- ——————————————— (1997) : Changes in the morphology of sediment accumulation in Shrivardhan Tidal bay, Ind. Jr. Of Geomorphology 2 (1) Academic and Law Serials New Delhi pp 51-56.

- ——————————————— (1998) : Reconstruction of the palaeo environment of Uran tidal flats, Maharashtra, in Ind. Jr. Of Geomorphology 2 (2) Academic and Law Serials, New Delhi pp 195-208.

- **Karlekar Shrikant, Gadkari Deepali (1998) :** Fossil littoral deposits of beach and dune origin at Nandgaon (Coastal Maharashtra), in, Ind. Jr. Of Geomorphology 3 (2) Academic and Law Serials, New Delhi pp 173-190.

- ——————————————— (1999) : Growth of mangroves on Padale inter tidal shore platform and falling sea level on Konkan coast of Maharashtra, in Maharashtra Bhoogolshastra samshodhan patrika 13(2), Pune, pp 95-104.

- ——————————————— (2000) : Kharlands of Mhasala creek, Maharashtra–A Geomorphic assessment, in Geomorphology and Remote sensing ed. V. C. Jha, acb publ. Calcuttai pp 173-183.

- ——————————————— (2000) : Response of Kelshi creek to a rising Sea level on Maharashtra coast as indicated by the satellite image, in Applications of remote sensing techniques, A special issue, MAEER'S MIT Pune Jr pp 43-47.

- ——————————————— (2000) : The evidences of the vertical displacement of shorelines in Konkan, (West coast of India), in, Quaternary Sea level variation, Shoreline displacement and coastal environments, Ed. G. Victor Rajamanickam, New Academic Publishers, Delhi.

- **Karlekar Shrikant, Deswandikar A. K (2000)** : Holocene Tidal basins of Diveagar and Valvati, Maharashtra coast in, Quaternary Sea level variation, Shoreline displacement and coastal environments, Ed. G. Victor Rajamanickam, New Academic Publishers, Delhi.

- **Karlekar Shrikant (2000)** : Tidal control on the survival of mangroves in disturbed coastal habitats of Kelshi creek, Maharashtra, in Ind. Jr. of Geomorphology 5 (1&2) Academic and Law Serials, New Delhi pp 63-70.

- ———————————————— **(2000)** : A geomorphic assessment of the potential and prospect of aquaculture in Majgaon creek on Konkan coast, in Ind. Jr. of Geomorphology 5 (1&2) Academic and Law Serials, New Delhi pp 161-167.

- ———————————————— **(2001)** : Placer enriched tidal sectors of streams around Purnagad creek, Maharashtra, India, in Hand book of placer mineral deposits Ed. G. Victor Rajamanickam, New Academic Publishers, Delhi 92. pp1-6.

- ———————————————— **(2001)** : Changing seasonal pattern of sedimentation in Kelshi creek and its environmental impact. MBP, Vol XV No. 2, July-December 2001 pp 95-104.

- ———————————————— **(2001)** : Assessing changein the coastal configuration and the sediment deposition, using image analysis technique (A case study of Kolamb Creek, Malvan, and Maharashtra) in Ind. Jr. Of Geomorphology 6 (1&2) Academic and Law Serials, New Delhi pp 75-82.

- ———————————————— **(2002)** : Geomorphology of the Konkan Coast, In Gography of Maharashtra, Ed. J. Diddee, S. R. Jog, V. S. Kale & V. S. Datye, Rawat Publications, Jaipur & New Delhi, PP 58-71.

- ———————————————— **(2002)** : Problem of beach erosion at Devbag, West coast of India, In Recent advances in Geomorphology, Quaternary Geology and Environmental Geosciences: Indian Case studies, Ed. S. K. Tandon & B. Thakur, Manisha Publications, New Delhi, PP 317-322.

- ———————————————— **(2002)** : Depositional tendencies on Velneshwar coast in Maharashtra Bhoogolshastra Sanshodhan patrika, MBP, Vol XVI No. 2 pp105-116.

- ———————————————— **(2003)** : Holocene fossil sedimentary feature of beach and dune origin on Varavade beach, Konkan, Maharashtra. In Ind. Jr. Of Geomorphology 8 (1&2) Academic and Law Serials, New Delhi pp129-134.

- ———————————————— **(2004)** : The impact of coastal landforms on the distribution of placer minerals along Maharashtra coast in Ind. Jr. Of Geomorphology 9 (1&2) Academic and Law Serials, New Delhi pp 101-114.

- ———————————————— **(2005)** : GIS application in the identification of reclamation strategies in Kharland development, Quaternary climatic changes and landforms edited by, N. Chandrasekar, M. S. Univ. Publ., pp 273-283.

- ———————————————— **(2007)** : Fossil sedimentary deposits in Ind. Jr. Of Geomorphology 11 (1&2) Academic and Law Serials, New Delhi pp101-114.

- ———————————————— **(2008)** : Use of directional derivatives in the study of estuarine sedimentation process on Konkan coast of Maharashtra. Trans Inst. Ind. Geographer 30 (2), pp 157-164.

- ——————————————— **(2009)** : Coastal processes and Landforms Diamond publication Pune, p-254.

- ——————————————— **(2009)** : Progress of researches in coastal Geomorphology in India, in Geomorphology in India: Prof. Savindra Singh felicitation volume, Prayag Pustak Bhavan, Allahabad, pp 217-228.

- ——————————————— **(2009)** : Coastal detrital latertites of Konkan Coast, Maharashtra, Co author, Thakurdesai S., in Geomorphology in India: Prof. Savindra Singh felicitation volume, Prayag Pustak Bhavan, Allahabad, pp 327 -340.

- ——————————————— **(2010)** : Coastal change and coastal area protection in Konkan, Maharashtra, The Deccan Geographer, vol. 48 No. 2, pp 85-94.

- ——————————————— **(2011)** : Closure of tidal inlets in a wave dominated, micro tidal environment along south Konkan coast of Maharashtra, Trans Inst. Ind. Geographer 33 (1), pp 38-42.

- ——————————————— **(2012)** : Late Holocene Geomorphology of Konkan coast of Maharashtra, Trans Inst. Ind. Geographer 34 (1), pp 21-34.

- ——————————————— **(2012)** : Vegetation response to South Asian Monsoon variations in Konkan, western India During the Late Quaternary: Evidence from fluvio-lacustrine archives, Co-authors K. P. N. Kumaran, Ruta B. Limaye, Sachin A. Punekar, S. N. Rajaguru, S. V Joshi, Quaternary International xxx (2012) 1 to 16,

- ——-—————————— **(2012)** : Landslide hazard zonation in Raigad district of Maharashtra: A multivariate approach, Jr. of Indian Geomorphology 1 acb publ. Kolkata, India pp 75-82.

- **Karlekar Shrikant, Shitole Tushar- (2013)** : Morphology and planation of intertidal rock platforms produced by water layer leveling on Dahanu coast of Maharashtra, Trans Inst. Ind. Geographer 35 (2), pp 239-248.

- **Karlekar Shrikant (2013)** : Coastal change: Case studies from Konkan, Maharashtra, in 'Landforms processes and environment management' Prof M. K. Bandyopadhyay felicitation volume, acb publ. Kolkata, pp 271-281.

- ——————————————— **(2014)** : Beaches and beach systems on Maharashtra Coast, in the 'Proceedings of the National conference on Modern trends in Coastal and Estuarine Studies', A Tilak Maharashtra Vidyapeeth publ. pp 19-34.

- **Karlekar Shrikant, Keskar Umesh (2014)** : Geo environmental and ecological consequences rising sea level in Ucheli creek, Maharashtra, in 'Proceedings of the National conference on Modern trends in Coastal and Estuarine Studies', A Tilak Maharashtra Vidyapeeth publ, pp 107-120.

- **Karlekar Shrikant, Thakurdesai Surendra (2014)** : Reconstruction of palaeo shoreline using carbonaceous clay deposits near Ratnagiri, in 'Proceedings of the National conference on Modern trends in Coastal and Estuarine Studies', A Tilak Maharashtra Vidyapeeth publ, pp 135-152.

- **Karlekar Shrikant (2015)** : Beach response to natural headlands on South Konkan and Goa coat, Trans Inst. Ind. Geographer 37 (2), pp 201-211.

- ——————————————— **(2015)** : Shaped by Nature, Geomorphology of the Konkan coast, in Maharashtra Unlimited, Maharashtra Tourism Development Corporation Publ., Konkan special issue, 4(4), pp 14-19.

- **Keskar Umesh (1996) :** A Geomorphic study of sedimentation and sea-level variations in Banganga Estuary, Maharashtra. Unpublished Ph. D. Thisis University of Pune, Pune.

- **Klein D. V. (1976) :** Holocene tidal sedimentation Benchmark papers in Geology (30), Hutchinson & Ross, Pennsylvania.

- **Krishnamurthy, K. (1991) :** Coastal Zone Management (in Tamil Nadu State, India), (eds Natarajan, R., Dwivedi, S. N. and Rama chandran, S.), Ocean Data Centre, Anna University, pp. 253-257.

- **Krishnan M. S. (1982) :** Geology of India and Burma, CBS publ. Delhi, p 604

- **Krishnan M. S. and Roy, B. C. (1945)** : Titanium, Records, Geol. Survey of India Vol. 78, pp 1-32.

- **Krishnan S. (2001) :** Presidential address. International seminar on placer deposits. In, a handbook of placer mineral deposits, Ed Rajamanickam G. V. New academic publishers Delhi, pp XXVII-XXX.

- **Kudale, M. D., Kanetkar, C. N. and Poonawala, I. Z. (2004) :** Design wave prediction along the coast of India. In Proceedings of the 3rd Indian National Conference on Harbour and Ocean Engineering, NIO, Goa, vol. 1, pp. 31–38.

- **Kulkarni Purva (1993) :** Mangrove swamps at Kelshi, in Coastal geomorphology of Konkan, Ed, Shrikant Karlekar, Aparna publ Pune, p 291-315.

- **Kulkarni V. N. (1985) :** Geology of Gujarat, in 'Navanirman', Spl issue by irrigation R and B dept., Gujarat state Vol 2.

- **Kumar, V. S., Pathak, K. C., Pednekar, P., Raju, N. S. N., Gowthaman, R. (2006) :** Coastal processes along the Indian coastline. Current Science 91, 530–536.

- **Kumar, A., Jayappa K. S., (2009) :** Long and short-term shoreline changes along Mangalore coast. International Journal of Environmental Science and Technology 3, 177-188.

- **Kumar, A., Narayana A. C., Jayappa, K. S. (2010) :** Shoreline changes and morphology of spits along southern Karnataka, west coast of India : a remote sensing and statistics-based approach. Geomorphology 120, 133-152.

- **Kumar, V. S., Dora, G. U., Philip, S., Pednekar, P., Singh, J. (2011) :** Variations in tidal constituents along the nearshore waters of Karnataka, west coast of India. Journal of Coastal Research. 27, 824-829.

- **Kumar, V. S., Johnson G., Dora G. U., Chempalayil S. P., Singh J. Pednekar P. (2012) :** Variations in nearshore waves along Karnataka, west coast of India. Journal of Earth System Science121, 393-403.

- **Kumar A., Seralathan P., Jayappa K. S. (2009) :** Distribution of coastal cliffs in Kerala, India, their mechanisms of failure and related human engineering response, in Environmental Geology, Vol. 58, pp 815.

- **Kumar Mohir, Chauhan H. B., Rajawat A. S. and Ajai (2012) :** Application of remote sensing and GIS techniques in understanding changes in mangrove cover in parts of Indus delta around Kori creek, Gujarat, India in Jr. of envir. Research and development, 7(1A), pp 504-511.

- **Kunte Pravin and Wagle B. G. (2001)** : Littoral transport studies along west coast of India- A Review, in Ind. Jr. of Marine Sci., Vol 30, pp 57 -64.

- **Kunte Pravin D., Wagle B. G. (2004)** : The beach ridges of India; A Review, in Jr. of Coastal research, spl issue No. 42.

- **La Fond, E. C. and Prasada Rao, R. (1953)** : Studies on sand movement across the Waltair Beach. *Curr. Sci.*, 1953, 22, 264-265.

- **Loveson V. J. (1993)** : Geological and Geomophological investigation related to sea-level variation and heavy mineral accumulation along the southern Tamilnadu beaches, India. Ph. D. thesis, Madurai Kamaraj University, 223 p.

- **Loveson V. J., Rajamanickam V. (1987)** : Coastal geomorphology of southern Tamil Nadu, India, in Proc. Nat. symp. on land transformation and management, Hyderabad (eds), Bhan and Zha, Ind. Soc. of remote sensing, Dehradun, pp 115-129.

- **Loveson, V. J., Rajamanickam, V. G. and Chandrasekekar, N. (1990)** : in Sea Level Variation and its Impact on Coastal Environment (ed. Rajamanickam, V. G.), 1990, pp. 159-178.

- **Mcmanus, J. (1988)** : Grain size determination and interpretation. In: Techniques in Sedimentology, Tucker M. (ed.). Blackwell: Oxford; 63-85.

- **Mahalik N. K., Das C., Maejima Wataru (1996)** : Geomorphology and evolution of Mhanadi Delta, India, in Jr. of Geosciences, Osaka city Univ., 39, Art. 6, pp 111-122.

- **Mahapatra Manik, Ratheesh R., Rajawat A. S. (2013)** : Shoreline change analysis along the coast of South Gujarat, India using remote sensing and GIS techniques, in Int. Jr. of Geology, Earth and Environmental sciences 3(2), pp 115 -120.

- **Mahapatra Manik, Ratheesh R., Rajawat A. S. (2014)** : Shoreline change analysis along the coast of South Gujarat, India using digital shoreline analysis system, in Jr. of Indian Soc. Of remote sensing 42(4), pp 869-876.

- **Majethiya Heman V., Bhatt Nishith Y., Patel Jagdish M., Solanki Paras M and Patel Satish J. (2014)** : Geoeomorphologicale study of the deltaic coast (west of kori creek) of kachchh, western india - using remote sensing and GIS International Symposium on "Operational Remote Sensing Applications: Opportunities, Progress and Challenges", Hyderabad, India.

- **Malik T. K. (1972)** : Opaque minerals from shelf sediments off Mangalore, West coast of India, Marine Geology, 12 : pp 203-211.

- **Mallik, T. K. (1983)** : Indian J. Mar. Sci., 12, 203-208.

- **Malik Javed N., C. V. R. Murty, M. EERI, and Durgesh C. Rai (2006)** : Landscape Changes in the Andaman and Nicobar Islands (India) after the December 2004 Great Sumatra Earthquake and Indian Ocean Tsunami, in Earthquake Spectra, Volume 22, No. S3, pages S43-S66, , Earthquake Engineering Research Institute, Kanpur.

- **Mane R. B. and Gowade M. K. (1974)** : Report on the prospecting of ilmenite beach sands of Ratnagiri district, unpubl. report, director of Geol. And mines, Govt. of Maharashtra, Nagpur.

- **Manjunath B. R., Balakrishna K. (1999)** : The depositional history of late quaternary sediments around Mangalore, west coast of India, in Ind. Jr. of Marine Sci. Vol. 28, pp 449-454.

- **Marathe A. R. (1981)** : Geoarchaeology of Hiran valley, Saurashtra, india, , Deccan college post graduate and research institute, Pune Publication.

- **Marathe A. R., Rajaguru S. N. and Lele V. S. (1977)** : On the problem of origin and age of miliolite rocks of the hiran valley, Saurashtra, India, in Sedimentory Geology 19 (197-215).

- **Mascarenhas Antonio (1998)** : Coastal sand dune ecosystems of Goa; significance, uses and anthropogenic impacts, in Study of Goa and its environment from space, NIO Report. 43 P.

- **Masselink Gerhard and Hughes M. G. (2003)** : Introduction to coastal process and Geomorphology, Arhold, London.

- **McCave I. N. (1982)** : Deposition of fine grained suspended sediments from tidal currents. Erosion and sediment yield. Bench mark papers in Geol. Vol. 63. Hutchinson Ross, Pennsylvania.

- **Meijerink A. M. (1971)** : Reconnaissance survey of the Quaternary Geology of the Caurvey delta. Journ. Geol. Soc. India 12, pp. 113-124.

- **Merh S. S. (1992)** : Quaternary sea level changes along Indian coast, in Proc. Indian natn. Sci. Acad., 58, A, No. 5, 1992, pp. 461-472.

- **Milovsky A. V. and Kononov O. V. (1985)** : Minerology, Mir Publishers, Moscow, p 320.

- **Mohan P. M. (1997)** : Distribution of heavy minerals in Parangipatti beach, Tamilnadu, Jr. Geol. Soc. Of India, 46 (4) : pp 401-408.

- **Mohan P. M. and Rajamanickam G. V. (2000)** : Buried placer mineral deposits along the east coast between Chennai and Pondicherry, Jr. Geol. Soc. Of India, Vol. 56 pp 1-3.

- **Mohan P. M. and Rajamanickam G. V. (2001)** : Indian Beach placers- A Review, in A handbook of placer mineral deposits, Ed Rajamanickam G. V. New academic publishers Delhi, pp 23-52.

- **Mohanti Manmohan and Swain Manas Ranjan (2005)** : Mahanadi river delta, east coast of India : an overview on evolution and dynamic processes, Department of Geology, Utakal university, Vani vihar, (Online) Available: http://www. megadelta. ecnu. edu. cn.

- **Mohr A. W. (1982)** : Sediment Control through dredging. Estuaries comparison ed. V. S. Kennedy Academic press, New York.

- **Moreno(1999)** : Morphology and Evolution of Landforms. Dept. of Geology, University of Delhi, New Delhi, pp. 121-130.

- **Mukherjee, A. D. and Chatterjee, S. (1997)** : Coastal erosion and accretion at and around Digha in Medinipur District of West Bengal. Indian Journal of Geography and Environment. 2. pp. 1-11.

- **Mukherjee T. K. (2001)** : Exploitation of heavy mineral resources in India, Inaugural address in A handbook of placer mineral deposits, Ed Rajamanickam G. V. New academic publishers Delhi, pp IX-XXVI.

- **Mukhopadhyay R. and Karisiddaih (2014)** : The Indian coastline: Processes and Landforms, in Landscapes and Landforms of India (ed)Vishwas Kale, Springer, London, pp91-101.

- **Nageswara Rao K, Sadakata N (1993) :** Holocene evolution of deltas on the east coast of India. In: Kay R (ed) Deltas of the World. ASCE, New York, pp 1-15.

- **Nagamalleswara Rao (1998) :** Density, mineralogy and textural studies of the beach and dune placer deposits of Andhra Pradesh coast, India, Jr. of Ind. Asso. of Sedimentologists, 17(2): pp 157-165.

- **Nageswara Rao, K and R. Vaidyanadhan (1978) :** Geomorphic features in Krishna delta and its evolution. Symposium on Morphology and Evolution of Landforms. Dept. of Geology, University of Delhi, New Delhi, pp. 121-130.

- **Nageswara Rao, K and R. Vaidyanadhan (1978) :** Geomorphic features in Krishna delta and its evolution. Symposium on Morphology and Evolution of Landforms. Dept. of Geology, University of Delhi, New Delhi, pp. 121-130.

- **Nageswara Rao, K. (1985a) :** Evolution and dynamics of the Krishna delta, India, in National Geographical Jour. India 31: 1-9.

- **Nageswara Rao, K. (1985b) :** Evolution of landforms in the area between the Krishna and Godavari deltas, in Indian Geographical Journal 60: 30-36.

- **Nageswara Rao K., Subraelu P., Venkateswara Rao. T., Hema Malini B., Rateesh R., Bhattachrya s., Rajwat A. S., Ajai (2008) :** sea level rise and coastal vulnerability: An assessment of Andhra Pradesh Coast, India, through remote sensing and GIS, in Jr. of Coast conservation (12), pp 195-207.

- **Nageswara Rao K, Sadakata N, Hema Malini B, Takayasu K (2005) :** Sedimentation processes and asymmetric development of the Godavari delta, India, In: Gioson L, Bhattacharya JP (ed). River Deltas. SEPM Spl Publ 83:435-45.

- **Nageswara Rao K, Ashokvardhan D, Subraelu P (2007) :** Coastal Topography and Tsunami Impact: GIS/GPS-based Mobile Map-ping of the coastal sectors affected by 2004-tsunami in Krishna– Godavari delta region. East Geogr 13:67-74.

- **Nageswara Rao K, Subraelu P, Rajawat AS, Ajai (2008) :** Beach erosion in Visakhapatnam: causes and remedies. East Geogr 14:1-8.

- **Naik Prabir kumar, Hota Ravindra Nath (2014) :** Geomorphological study of sand dunes with special reference to their hydrogeology in southern coast of Odisha, India, in Int. research Jr. of Earth Sci. 2(9), pp 15-21.

- **Nair M. M. (1987) :** Coastal Geomorphology of Kerala, in Jr. of Geol. Soc. of India, 29 (4).

- **Nair, R. R., Hashimi, N. H. and Rao, P. C. (1982) :** Mar. Geol., 47, 77–86. Nair and Rao (1991).

- **Narayan and Anirudhan(2003) :** Nature, distribution origin and economic evaluation, Ph. D. thesis, Unpubl., Tamil university, Thanjavur, P 0.

- **Nayak, B. U. (1980) :** Coastal erosion in India. Causes, processes and re-medial measures. In Proceedings of GEOTECH-80, Conference on Geotechnical Engineering, IIT, Bombay, vol 2, pp. 55-63.

- **Nayak G. N. & Chavadi V. C. (1989) :** Distribution of heavy minerals in beach sediments around kali river, Karwar, Geol. Sur. of India, Spl. Publ. 24: pp 241 –245

- **Nayak G. N. (1993) :** Beaches of Karwar, Rajhans Vitaran, Panji, Goa, Publ No. 108.

- **Nayak G. N. (1997) :** Grain size parameters and heavy minerals as indicators of depositional environment – A case study of beach sediments from Goa, Jr. of Ind. Asso. Of Sedimentologists, 16(1): pp 127-139.

- **Nayak G. N. (2005)** : The Indian ocean coastline, Coastal Geomorphology, in Encyclopedia of coastal science (ed) Schwartz Maurice, Springer, , London, pp 555-556.

- **NIO report on Rewas (1985)** : Analysis of water and bed samples collected from Mora and Rewas Port by National Institute of Oceanography. For release of waste water from Nogothane petrochemical complex of IPCL.

- **Niyogi D. (1971)** : Studies in Earth Science (West Comm Vol Today &Tomorrow), p 289.

- **P. Chandramohan, Jena B. K. and V. Sanil Kumar (2001)** : Littoral drift sources and sinks along the Indian coast, in Current science, VOL. 81, NO. 3, PP 292 – 297.

- **Pachauri, R. K. (1996)** : Changing Coastlines. Effects of Climatic Change. Tata Energy Research Institute, Mumbai. pp 11-20.

- **Parthasarathi Chakravarti (1995)** : Evolutionary history of coastal quaternaries of the Bengal plain, India, in Proc. Nat. Sci. and Acad. 61, A, No. 5, pp 343-354.

- **Paul A. K. (2002)** : Coastal Geomorphology and Environment, ACB Publications, 575 pp.

- **Paul, A. K., Mukherjee, C. N., Bhattacharyya, P. and Mazumder (2004)** : Environmental management and prospects of coastal tourism under the changing shoreline characters of Hugli-Subarnarekha complex. Ind. Jr. of Geography and Environment. 8 & 9. pp. 1-27.

- **Pethic Jhon, (1984)** An introduction to coastal geomorphology, Arnold – Heinmann Publishers (India) Pvt. Ltd., New Delhi, pp 127-143.

- **Pethick, J., (1992)** : Natural Change, In: Coastal zone planning and management, Proceedings of the conference Coastal management '92, organized by the Institution of Civil Engineers and held in Blackpool on 11-13 May 1992, Thomas Telford, London. pp. 49-63.

- **Petro B. C. (1989)** : A study of heavy mineral sands of Gopalpur beach, Orissa, Jr. of Geol. Soc. Of India 33: pp 243-252.

- **Prabakaran K., Anbarasu K. (2010)** : Coastal Geomorphology and evolution of Rameshwaram island, Tamil Nadu. India, in Research Jr. of Earth Sci. 2 (2), pp 30-35.

- **Prakash Vinu (2012)** : Sedimentological and Geotechnical studies of coastal sediments of central Kerala; A Ph. D. thesis, Dept of Marine Geology and Geophysics: School of Marine sciences, Cochin univ. of Sci. and Tech. pp 19-24.

- **Prudhvi Raju, K. N. & Vaidyanadhan, R. (1978)** : Geomorphology of Visakhapatnam, Andhra Pradesh, J. Geological Society of India, 19, 26-34.

- **Psuty (1992)** : Spatial variation in coastal foredune development in coastal Dunes by RWG.

- **Raghavan B. R., Vinod B. T., Dimple K. A., H. Venkatesh Prabhu, Udayshankar H. N., Shreedhara Murthy T. R., (2001)** : Evaluation of the Nethravati spit complex, West coast of India: Integrated change detection study using topographic and remotely sensed data, in Ind. Jr. of Marine Sci. 30 (4), pp 268-270.

- **Rajamanickam G. Victor & Gujar A. R. (1995) :** Transparent heavy mineral distribution in the bays of Jaygad, Ambawali and Varvade Ratnagiri Dist. Maharashtra, Jr. of Ind Assso. Of Sedimentologists 14:pp 43-53.

- **Rajamanickam, G. V. (1990) :** Sea Level Variation and its Impact on Coastal Environment. Tamil University publication No. 131. Tamilnadu. pp xiii-xvi.

- **Rajamanickam G. Victor (2001) :** A handbook of placer mineral deposits, (Ed) New academic publishers Delhi, p 327.

- **Ramachandran, S., and Natarajan, R. (1990) :** Impact of Sea Level Rise on Estuarine Ecosystems, In : Sea level variation and its impact on coastal environment Ed Rajamanickam, G. V., 1990. Tamil Universtiy, Thanjavur. pp 267-278.

- **Ramasamy (1991) :** A remote sensing study of river deltas of Tamil Nadu, in Memoirs, Geol. Soc. of India Vol 22, pp 75-89.

- **Ramasamy S. (1996) :** Heavy mineral distribution around Mahavala Kuruchi Kanyakumari, Tamilnadu, Jr. of Ind. Asso. of Sedimentologists 15:pp 29 –42.

- **Ramesh R., Ramachandran P., Senthil Vel A. (2011) :** National assessment of shoreline change, Puduchery Coast, NCSCM/MoEF report, p 57.

- **Ramkumar M. (2000) :** Recent changes in the Kakinada Spit, Godavari delta, in Jr. of Geol. Soc. of India, Vol. 55, pp 184 -187.

- **Ranjana and Anjum Farooqui (2015) :** Relative Sea level, Climate and Geomorphological changes since 8 ka in Krishna river delta, India, in Int. J. Curr. Res. Aca. Rev. 2015; 3(10): 303-316.

- **Ranwell (1972) :** Ecology of salt marshes and sand dunes, Chapman and Hall, London pp 63-95

- **Rao T. R., Rao N. R. and Rama Raju M. V. (1983) :** Ilmenite of black sand deposits of Visakhapatnam-Bhiminipatnam beach, East coast of India, Indian Jr. of Mar. Sci. Vol. 12 pp 220-222.

- **Rao D. G. (1988) :** Mar. Geol. 82, 277-283.

- **Rao, K. L. (1979) :** India's Water Wealth: Its Assessment, Uses and Projections Orient Longman Ltd, New Delhi, p. 267.

- **Reading H. G. (1986) :** Sedimentary environments and facies Blackwell Scientific, Oxford, London.

- **Roberts Neil (1989) :** The Holocene-An environmental history, Basil Blackwell Oxford.

- **Saha sourav, Burley S. D., Banerjee S., Ghosh Anupam, Saraswati P. K. (2016) :** The morphology and evolution of tidal sand bodies in the macrotidal gulf of Khambat, western india, , in Marine and Petroleum Geology, Elsvier, pp 1-17.

- **Sajeev, R. (1993) :** Beach dynamics of Kerala coast in relation to land–sea interaction. Ph D thesis, Cochin University of Science and Tech-nology, p. 130.

- **Sakurkar, C. (1999) :** Contributions of Stratigraphy and Palynology of the Quaternary Sediments of Maharashtra and Goa, India. Summary of unpubl. Ph. D. thesis in Geol., Univ. of Pune.

- **Sambasiva Rao, M. & Vaidyanadhan, R. (1979) :** Morphology and evolution of Godavari delta. – Zeitschrift fur Geomorphologie N. F. 23: 243-255.

- **Sambasiva Rao M (1982) :** Morphology and Evolution of the modern Cauvery delta,

Tamil Nadu India. Trans. Inst. Indian Geographers Vol. 4 No. 1, pp 67-78.

- **Sambasiva Rao, M. & Vaidyanadhan, R. (1979)** : Morphology and evolution of Godavari.

- **Samsuddin, M. and Suchindan, G. K. (1987)** : Beach erosion and accretion in relation to seasonal longshore current variation in the northern Kerala coast, India. J. Coast. Res., 3, 55-62.

- **Samsuddin, M., Ramachandran, K. K. and Suchindan, G. K. (1991**) : Sediment characteristics, processes and stability of the beaches in the northern Kerala cost, India. J. Geol. Soc. India, 38, 82-95.

- **Sastry A. V. R., Swamy A. S. R. and Vasudev K. (1987)** : Heavy minerals of beach sands along Visakhapatnam-Bhiminipatnam beach, Ind Jr of Mar. Sci. Vol. 16 pp 39-42.

- **Sastry A. V. R., Swamy A. S R. and Rao N. V. N. D. P. (1981)** : Distribution of Garnet sands along Visakhapatnam-Bhiminipatnam beach, Ind Jr of Mar. Sci. Vol. 10 pp 369-370.

- **Sathasivam Sathish., Kankara R. S., Chenthanil Selvan S., Muthusamy Manikandan, Samykannu Arockiraj, Bhoopathi Rajan (2015)** : Textural characterization of coastal sediments along Tamil Nadu coast, East coast of India, in Procedia engineering, Vol. 116, pp 794-801.

- **Selley R. C. (1976)** : Ancient Sedimentary Environments Chapman and Hall, London

- **Selvam V., Gnanappazham I., Navamuniyammal M., Ravichandran K. K., Karunagaran V. M. (2002)** : Atlas of Mangrove Wetlands of India, Part I, Tamil Nadu: M. S. Swaminathan research foundation, Chennai, pp 5 -13.

- **Sherman & Baner (1993)** : Dynamics of beach dune systems, in Progress in physical Geog.

- **Shetye S R, Gouveia A D, Singbal S Y, Naik C G, Sun-dar D, Michael G S and Nampoothiri G (1995)** : Propa-gation of tides in the Mandovi-Zuari estuarine network. Proc. Indian Acad. Sci. (Earth Planet. Sci.) 104(4), pp 667-682.

- **Shitole (1995)** : Surf zone dynamics of the wave dominated beach at Hedvi, Coastal geomorphology of Konkan, ed. Karlekar, Shrikant, Aparna publication, Pune. pp 7-36.

- **Short A. D. (2000)** : Beach and shoreface morphodynamics: School of Geosciences, Univ. of Sydney publ, p 217.

- **Short A. D. & Hesp (1982)** : Wave, beach and dune interactions in South eastern Australia, Marine Geology 48.

- **Short, A. D. (1984)** : Beach and nearshore facies: Southeast Australia. Mar. Geol. 60, 261–282.

- **Shrikhande Bhagyashree (1993)** : The depositional dynamics of the intertidal spit bar at Rewas. Coastal geomorphology of Konkan ed. Shrikant Karlekar, Aparna Publicaitons (PP 132-156).

- **Shrikhande Bhagyashree (1994)** : The pattern of sedimentation in Dharmatar Creek and its impact on the development of mud beach at Revas Unpublished Ph. D. Thesis, University of Pune, Pune.

- **Siddiquie & Rajamanickam G. V. (1979)** : Offshore ilmenite placer of Ratnagiri, Konkan coast, Maharashtra, Marine mining, 2 : pp 29-118.

- **Singh Indra Bir (1990)** : Sedimentary Structures : Methods of Study and Significance in Coastal Sediments. Workshop on Coastal Geomorphology, Andhra University, Vishakhapatnam (pp 23-27).

- **Smirnov V. I. (1976)** : Geology of mineral deposits. MIR publishers, pp 390-425.

- **Smith D. (1984)** : The hydrology and geomorphology of tidal basins in Closure of Tidal basins, Delft University press, Netherlands.

- **Somanna k. Somasekara Reddy T., Sambasiv Rao M. (2016)** : Geomorphology and evolution of the modern Mahanadi Delta using remote sensing data, in Int. Jr. of Sci. and research, 5(2), pp 1331-1333.

- **Srihar V. A., Modi V. K., Mane R. V., Deolac C. B., Deshmukh d. g. and Bhandury S. K. (1991)** : Geology and mineral resources of Maharashtra, Directorate of Geology and mining, Govt. of Maharashtra, Nagpur, R. No. G 220, P 365.

- **Srinivas Kumar T., Mahendra R. S., Nayak Shailash, Radhakrishnan K., Sahu K. C. (2010)** : Coastal vulnerability assessment for Orissa state, East coast of India, in Jr. of Coastal research, 26 (3), pp 527-530.

- **Srinivasan R. (2001)** : Placer mineral deposits-Need for a pragmatic approach, Key note Address in, A handbook of placer mineral deposits, (Ed) New academic publishers Delhi, ppXXI-XXXVI.

- **Stoddart D. R. (1973)** : Coral reefs of the Indian Ocean, In: Biology and Geology of Coral Reefs, (Ed. Jones, O. A. & Endean. R.) Geology. 1, 51-92.

- **Stride A. H. (1982)** : Offshore tidal sands – processes and deposits Chapman and Hall, London.

- **Struaten L. M. J. U. (1964)** : Deltaic and shallow marine deposits, Dev. in Sed. Vol. 1. Elsevier Publishing, Amsterdam.

- **Subramanian, V. (1993)** : Curr. Sci., 64, 928-930.

- **Sukhtankar, R. K., Pandian, R. S. (1990)** : Evaluation of Sea Level Rise on the Shore Zone Areas of the Maharashtra Coast, Sea Level variation and its impact on coastal environment. Ed. Rajamanickam G. V., Tamil University Press, publication No. 131. Thanjavur, Tamilnadu, pp 329-337.

- **Sukhtankar, R. K., PawarJ. B., Kulkarni M. B., (1986)** : Quaternary Sediments in Relation to Geomorphology and Tectonics along the Vengurla cast, Maharashtra, Science and Cultivation, 52(3), pp 95-98.

- **Sundar D., Shetye S. R (2005):** Tides in the Mandovi and Zuari estuaries, Goa, west coast of India, in Jr. Of Earth Systems Sci 114 (5), pp 493-503.

- **Syriac S. Roy G. and Damodaran K. T. (1990)** : Studies on the distribution of organic matter and carbonate content of sediments in Mahe Estuary, North Kerala, in Jr. Geol. Soc. of India 36 (3), Bangalore.

- **Thakurdesai Surendra (2004)** : The geomorphic assessment and image appraisal of laterites and their effect on landform and landuse of Kolambe-Golap plateau, Maharashtra, unpublished Ph. D. Thesis, University of Pune.

- **Theenadhayalan G., Kanmani T., Bhaskaran R. (2012)** : Geomorphology of the Tamil Nadu coastal zone in India: Applications of Geospatial Technology, in Jr. of Coastal research, 28 (1), pp 149-160.

- **Thomas, Michael F. (1974)** : Tropical Geomorphology, The Macmillan press ltd., London.

- **Trenhaile A. S. (1987)** : Geomorphology of rock coasts, Clarendon press, London.

- **Trenhaile A. S. (1997)** : Coastal dynamics and landforms, Clarendon press, London.

- **Trivedy R. K. and Goel P. K. (1986)** : Chemical and biological methods for water pollution studies, Environmental publications, Karad.

- **Unnikrishnan a. s. (2010)** : Tidal progression off the central west coast of India, in Ind. Jr. of Geo-Marine sci., 39(4), pp 485-488.

- **Usha, N. and Subramanian, S. P. (1993)** : Seasonal shoreline oscillation of Tamil Nadu Coast. Curr. Sci., 65, 667-668.

- **Usha, N. and Subramanian, S. P. (1993)** : Curr. Sci., 65, 667–668.

- **V. Sanil Kumar, Pathak K. C., Pednekar P. Raju N. S. N. and Gowthaman R. (2006)** : coastal processes along the Indian coastline, in Current Science 91 (4), Research communication, pp 530-536.

- **Vaidyanadhan R., (1991)** : Status of Quaternary delta studies in India. *In* R. VAIDVANADHAN, ed. : Quaternary deltas in India. Mem. Geol. Soc. India, No. 22, 1-11.

- **Vaidyanadhan R., (1987)** : Coastal Geomorphology in India, Jr. Geol. Soc. of India, 29, pp 373-378.

- **Vinayaraj P., Glejin j., Udhaba dora G., Philip sanjiv C, Sanil Kumar V, Gowthaman r, (2011)** : Quantitative estimation of coastal change along selected locations of Karnataka, India, A GIS and RS approach, in Int. Jr. of Geosciences Vol. 2, pp 385-393.

- **Vivekanandan V. and Kurien John (1980)** : Aquaculture- Where greed overrides need, in The Hindu Survey of the environment, Chennai pp 27-33.

- **Vora, K. H., Chauhan, O. S. and Rao, B. R.,** *Indian J. Mar. Sci.,* 1987, 16, 230–234.

- **Wafar W., Wafar S., Yennavar P., (2005)** : The Indian ocean coastline, Coastal Geomorphology, in Encyclopedia of coastal science (ed) Schwartz Maurice, Springer, , London, pp 556-564.

- **Wafar W., Wafar S., Yennavar P., (2005)** : The Indian ocean coastline, Coastal Geomorphology, in Encyclopedia of coastal science (ed) Schwartz Maurice, Springer, London, pp 556-564.

- **Wafar, M. V. M. (1986)** : Corals and Coral reefs of India, Proc. Indian Academy of Sciences, Suppl. pp 19-43.

- **Wagle, B. G. (1987)** : Ph D thesis, University of Mumbai, p. 166.

- **Wagle B. G., Gujar A. R. & Mislankar P. G. (1989)** : Impact of coastal features on beach placers: A case study using remote sensing data, Offshore technology conference, publ 6065, pp 229-233.

- **Wagle B. G (1993)** : Geomorphology of Goa and Goa Coast. A review, in Giornale di Geologia, ser. 3, vol. 5512, pp. 19-24, Bologna.

- **Wood E. M. (1983)** : Corals of the World. (T. F. H. Publications. Inc., Ltd.).

- **Woodroffe Colin D., (2002)** : Coasts: form, process and evolution, Cambridge Univeersity Press, U. K.

INDEX

ABOUT THE AUTHOR

Dr. Shrikant Karlekar

Dr. Shrikant Karlekar (M.Sc.Ph.D.) former Head Department of Geography, Sir Parashurambhau College, Pune, Maharashtra and Dean faculty of Earth Sciences, Tilak Maharashtra Vidyapeeth, Pune, has worked extensively in the field of Coastal Geomorphology since 1981.His doctoral thesis titled "A Geomorphic Study of South Konkan"was followed by about 80 research papers mainly in the field of Coastal Geomorphology. Most of his research scholars have worked for their M.Phil. and Ph.D. degrees on the problems and issues related to coastal areas. He has to his credit about 20 published books in field of Geomorphology, Remote Sensing, Statistical methods and GIS.